Forest Planning at Landscape Level

(A Case Study of Working Plan Revision in Chhattisgarh)

About the Author

Kailash Chandra Bebarta, an officer from Indian Forest Service, presently works in a senior position in forest department, Government of Chhattisgarh. Getting masters' degree in Zoology, he earned postgraduate diploma in Forestry from Indira Gandhi National Academy of Forestry, Dehradun. He also holds a diploma in business administration from Annamalai University. Mr. Bebarta has to his credit three books entitled, *Teak: Ecology, Silviculture, Management and Profitability; Planning for Forest Resources and Biodiversity Management* and *Forest Resources and Sustainable Development.* Also he prepared the working plan for the forests of Rajnandgaon district in the year 1998. Currently, he is engaged in developing fully functional Forest Information System (FIS) for the state of Chhattisgarh, using the state-of-the-art technologies like Geographic Information System (GIS), Remote Sensing (RS), Geoscience, and Information and Communication Technology (ICT) etc.

Forest Planning at Landscape Level

(A Case Study of Working Plan Revision in Chhattisgarh)

Kailash Chandra Bebarta

CONCEPT PUBLISHING COMPANY, NEW DELHI-110059

ISBN : 81-8069-369-4 (HB)

First Published 2007

Published and Printed by
Ashok Kumar Mittal
Concept Publishing Company
A/15-16, Commercial Block, Mohan Garden
NEW DELHI-110059 (India)

Phones : 25351460, 25351794
Fax : 091-11-25357103
Email : publishing@conceptpub.com

Preface

Forest working plan remains the basis of forest management practices since the last leg of twentieth century. During this period India heralded into an age of scientific management in forestry. But over the years, on account of a host of changes that occurred in social, economic, political, environmental landscape of the country as well as world, a paradigm shift has taken place in the concept and practices of forestry. Forestry principles and practices are continuously redefined on the basis of emerging ethos of sustainable development. The National Forest Policy, 1998 and the Rio Declaration have reiterated the need to have a management plan for every piece of forests. The Honourable Supreme Court of India, in its various decisions and directives, reaffirmed the compulsive need for a management plan for forests.

I wrote a book titled as *Planning for Forest Resources and Bio-Diversity Management* focusing on principles, organisation and methodologies involved in the process of working plan preparation. Till the publication of my book, in fact there was no comprehensive book dwelling on fundamental principles and methodology of preparing or revising forest-working plan. Preparing a working plan for a forest division necessitates a coordinated application of principles of forestry, management and social sciences. It is complex job, which needs synchronization and optimization of management objectives. Therefore, formulating or revising a forest-working plan has remained a challenge in the history of scientist forestry. Again, in the absence of any comprehensive book on how to prepare working plan, the task of revising working plan becomes more arduous and intractable. Therefore, the author attempted to prepare a case study on the processes of revising the forest-working plan of Rajnandgaon district, in Chhattisgarh.

I wrote this book to introduce foresters and students in forestry to the process of working plan revision/formulation. Till date there is, to my knowledge, no case study dwelling on the process of revising any working plan in India. I wanted to present different strands of thoughts to foresters and students who are given the job of formulating a working plan at a landscape level. All such strands intertwine in the fabric of sustainable forest management.

Rajnandgaon constitutes the study area. It is one of the revenue districts in the state of Chhattisgarh. Till the year 2000, it used to be one forest division. But after the formation of Chhattisgarh in the year 2000, the old forest division of Rajnandgaon falling in the revenue district of same name was bifurcated into two forest divisions: Rajnandgaon and Kheiragarh. It constitutes a landscape from the point of natural traits such as landform, soil characteristics, watershed attributes and forest resources.

In writing the book, I started with the premise that it is more useful for practicing foresters and students to learn about the past management practices and conditions that give rise to different strategies for sustainable forest management. The book is organised into seven chapters grouped into two parts. The first part deals with fundamental aspects of planning which are germane to formulating a pragmatic working plan. It dwells upon the concept of planning, types of plan, planning process, definition and function of working plan, changing role of working plan, objectives of plan, components of working plan, stages in working plan process and organization of working plan areas. The second part deals with the process and methodology adopted for revising/formulating the working plan for the study area. It includes the fieldworks, and their relevance in decision-making process of the plan.

The first chapter delves in defining principles, process and methodology involved in revising a forest working plan. Working plan is a tactical silvicultural plan. It encompasses an integrated and time scheduled package of silvicultural prescriptions to achieve objects of management. It is implemented at landscape or forest management unit (FMU) level. The time horizon for such a plan is generally ten years. It means the plan will be revised in every ten years. Changes in the forest crop are reflected in the revised plan. Evolving an appropriate package of silvicultural prescriptions is an uphill task. It is basically a forest planning at landscape level or forest management unit level. The task of planning shall be accomplished successfully if the planner has the requisite knowledge and experience in the art and science of forest management.

The second chapter delves in definition of landscape, history, location, land use pattern, factors of locality, demographic dynamics, social fabric, occupational patterns, disturbance regime, configuration of forest areas, rights and concessions bestowed on the local people and assessment of forest resources available in the study area. These aspects are required to assess the strength and opportunities of the area for which planning is undertaken. Planning is the precursor of all resource management, including forests and wildlife. Planning helps in identifying what must be done and how it must be done. In other words, planning sets the stages and takes control of other key management practices of the organization. It comprises several closely related steps and activities.

The third chapter attempts to review the performance of past working plans pertaining to the study area. The working plan is generally meant for a definite period such as 10 years or 15 years. At the end of that period, a revised plan will be at place. Therefore, revision of the plan commences before the plan ends. Evaluation is an integral part of revision process. It is an important management function. The impact of the last plan on different aspect of forests is critically evaluated and on the basis of this, new plan prescriptions are evolved. In this chapter, sincere attempt has been made to evaluate the results of past management. Beside this, evaluation at the working circle level has also been undertaken to find out the impact of management practices on forest ecosystem as well as on the crop. An attempt has also been made to look at the impact of last plan's prescription on a holistic basis and make a stratum-wise impact analysis.

The fourth chapter intends to discuss a broad spectrum of strategies of forest management practices. Strategy plays an impotent role in forest management. It translates the policies and projects into action on the ground. This redeems the objectives of management. A sustainable forest management strategy consists of broad direction for forest conservation, development, sustainable use, equity based benefit sharing, etc. The purpose of evolving strategies is not only to determine and communicate a picture of the kind of working, that an organization visualizes, but also to develop practical framework for acting at ground level. This chapter, while dealing with sustainable strategy option delves upon introductory subjects such as three different models of forestry, principles of forest management at landscape level, policy support for management, strategy formation for forest modelling and activities under it.

The fifth chapter delves upon the concept and nuances of silvicultural management. Managing forests on the principle of sustainable development transcends beyond the domain of technical forestry. It involves myriad issues like social, economic, political betterment of human condition, social equity, poverty reduction, gender roles, governance structure, etc. Such issues lie at the heart of sustainable development and use of forest resources. Silviculture, as element of forestry, has the daunting responsibility of providing the biological and technical options to achieve objectives of sustainable development. Therefore, the aspect of silvicultural management plays a vital role in developing a working plan. This chapter discusses various management options in the context of the forest resources of the study area and the process involving the constitution of the working circles. Objectives and characteristics of each working circle have also been dealt threadbare in the chapter. In the end, regulation and management of important miscellaneous subjects like rights and concessions, management of ecotone area etc. have been discussed.

The sixth chapter deals with different issues and case studies related to participation of people in forest management. It is presented in four sections. First section introduces the subject of participation with a focus on concept, definition and relevance. Second section deals with a discussion on practice of participation in the state. The third section deals with two specific context-driven case studies. The first case study was conducted in the year 1998 in three villages under Joint Forest Management. The purpose of the study was to assess the adaptation of people in fuelwood stress condition. The study tried to determine the degree of fuelwood stress in these villages and its impact on participatory management also. The second case study pertains to the appraisal of performance of Joint Forest Management (JFM) activities in the study area as a whole. The evaluation of working of JFM committees in the study area was done on the basis of field inspection as well as computer aided grading methodology. The fifth section deals with the attendant problems in operationalising joint forest management and difficulties in achieving participation in a hierarchical organization like forest department and the need to change the working for scaling up community participation in forest department.

The seventh chapter presents a strategy to stimulate on-farm tree cultivation in the district. On farm cultivation of trees is a well-entrenched component of the farming system in Chhattisgarh. Over the years, on account of a dynamic change in rural economy, it has emerged as a private economic resource having immense social security value. Beside this, it has also helped in sustainable development of the existing forest resources in the district Rajnandgaon. Farmers in the Chhattisgarh belt manage trees as an integral part of the farming system. They evolve suitable management practices in a given ecological and socio-economical settings The objective of such practices is to produce goods like small timber, poles, fuelwood, fodder and other saleable tree products and services like conservation of soil and moisture. More importantly, looking into the cultural entity of Chhattisgarh, growing of trees in grooves also caters to the religious and cultural needs of the rural folk. As the diversity of goods and services derived from trees becomes appreciated, a wider potential was postulated for 'agro-forestry'— as a tool for resource poor farmers to stabilize and improve their farm system. But what stimulates the rural folk to go for on-farm tree growing is a matter to ponder over. There is no single reason responsible for the emergence of this tree growing culture among the farmers. A dynamic change in the rural economy in the Chhattisgarh region, more especially in the districts of Durg, Rajnandgaon, Raipur, Bilaspur has spurred widespread on farm tree growing in the rural landscape.

A great many contributed to this book, often in ways that were not apparent at the time. It is not possible to name them all here, but some of the most important deserve special thanks. Dr. M.N. Tandon, the then Chief Conservator of Forests, Working Plan, Madhya Pradesh, Mr. D.S. Sharma, the then Conservator of Forests, Working Plan, Jabalpur deserve special mention. These two senior forest officers along with other senior colleagues were responsible for giving pragmatic shape to my thoughts and approaches relating to forest management. This apart, I extend my hearty thank to my colleagues in Indian Forest Service in Chhattisgarh and Madhya Pradesh where I commenced my journey towards learning dynamisms and nuances of forest management, for sharing their knowledge and experiences with me in various occasions. I value the cooperation extended by staff at my office in preparing some illustrations.

Kailash Chandra Bebarta

Raipur,
30-05-2006
(bebarta@indiatimes.com)

Contents

List of Table

List of Figures

List of Boxes

PART - I

Working Plan Revision : Study Area, Methodology and Past Management

PART — I

1

Forest Working Plan

Principles, Processes and Methodology

Working plan is a tactical silvicultural plan. It encompasses an integrated and time scheduled package of silvicultural prescriptions to achieve objects of management. It is implemented at landscape or forest management unit (FMU) level. The time horizon for such a plan is generally ten years. It means the plan will be revised in every ten years. Changes in the forest crop are reflected in the revised plan. Evolving an appropriate package of silvicultural prescriptions is an up hill task. The task of planning shall be accomplished successfully if the planner has the requisite knowledge and experience in the art and science of forest management. This chapter has been organized in two parts. The first part deals with fundamental aspects of planning which are germane to formulating a pragmatic working plan. It dwells upon the concept of planning, types of plan, planning process, definition and function of working plan, changing role of working plan, objectives of plan, components of working plan, stages in working plan process and organization of working plan areas. The second part deals with the methodology adopted for formulating the working plan for the study area. It includes the fieldworks, and their relevance in decision-making process of the plan.

PART I

Defining Planning

Planning is 'task number one' of forest management. It symbolizes the determination of course of action to achieve a desired result. It aims at setting objectives of an organization. It is like deciding in advance what to do, how to do, when to do and who is to do. It is also an institutional process. It is characterized as the process of thinking before acting. Planning is the precursor of all other managerial functions. In fact, planning is the foundation of forest management.

Planning is a way of organizational life. Every organization in today's world plans its activity. Every plan is unique to the planner as well as to the situation. Planning begins with a goal or objective or targeted outcome that an organization wishes to achieve. A plan is a means to achieve the said goal. From the point of development, planning is of two types. The first one is informal type and the second one is formal type. Informal planning is a product of intuitive and instinctive reaction to a situation. Formal planning is a structured, systematic and scientific process of investigation, assessment and action.

Planning is deciding in advance what is to be done. It involves the selection of objectives, policies, procedures and programmes from among the available alternatives. A plan is a pre-determined course of action to achieve a specified goal. In short, it is a blue print for action. Different authors, however, have defined planning differently. Few of such definitions are as follows :

1. David Cleland and William King: "Planning is the process of thinking through and making explicit the strategy, action and relationship necessary to accomplish an overall objective or purpose."
2. George R. Terry: "Planning is the selecting and relating of facts and the making and using of assumptions regarding the future in the visualization and formation of proposed activities believed necessary to achieve desired results."
3. Theo Haimann: "Planning is deciding in advance what is to be done. When a manager plans, he projects a course of action for the future, attempting to achieve a consistent, coordinated structure of operations aimed at the desired results."
4. Koontz O'Donnell: "Planning is an intellectual process, the conscious determination of courses of action, the basing of decisions on purpose, acts and considered estimates."
5. George B. Galloway: "Planning is the opposite of improvising. In simple words, it is an organized foresight plus corrective hindsight... Conceived as a process, planning embraces a series of steps: (i) the determination of objectives to be sought, (ii) research to understand the problem, (iii) the

discovery of alternative solutions, (iv) policy making and chosing between alternatives, including the frequent choice of doing nothing, and (v) the detailed execution of the chosen alternative known in physical planning as lay out or design."

6. Alfrod and Beatty : "Planning is the thinking process, the organised foresight, the vision based on facts and experience that is required for intelligent action."

Objectives of Forest Management Planning

Preparing a working plan necessitates planning. This type of planning is known as forest management planning. The objectives of forest management planning in general are as follows :

- To promote sustainable development of forest resources.
- To utilize forest resources and other natural endowments.
- To serve the village community and the society at large by producing the tangible and intangible benefits.
- To use the resources in such a way that it helps in maximizing the benefits to the society on a sustainable basis. Benefits must encompass goods and services including life supporting services.

Planning Approach

Planning, by simple definition, is to 'devise detailed methods for doing, arranging, and making something'. For doing different things different approach should be adopted. Some useful approaches employed in forest management planning are explained as follows :

1. Bottom-up Approach

This approach necessitates planning from the bottom or base level. The problems at the base level should be studied first and data at the bottom level should be collected. On the basis of the problems at the bottom level the plan should be evolved to address it. Therefore, the approach which looks at the bottom level for identification of problem and for generation of supporting data is called bottom-up approach. Many management plans failed because such an approach was not adopted in the past. Generally, local communities are not adequately involved in forest management planning process. As a result such plan miserably fails. Therefore this approach has immense relevance to the field of forest management planning. There are several ways to involve people in the planning process. The present practice of joint forest management is one of them. Involving villagers and communities for planning development of forests, pasture and protected areas constitutes bottom-up approach.

2. Iterative Approach

Planning basically is an iterative approach. Before a final plan is prepared, many surveys, studies, assessments, alternative considerations and revisions will have to be made. Generally in the field of forest management planning one or two preliminary reports are prepared on the basis of management need. Such reports are prepared by gathering, analyzing and interpreting data from the field. After a thorough discussion on such reports in the review committee, generally consisting of senior officers from different branches of forestry, an ultimate strategy for sustainable development of forests is evolved. Therefore, in the process of finalization of management plan many steps are repeated to hammer out the best appropriate strategy. Such steps may consist of different field study, analysis and presentation of reports.

3. Flexible Approach

An approach is called flexible when it leaves enough room for future adjustments, modification and revision. A management plan is the starting point of forest management. The plan is drawn for a period of 10 years. On account of long span of the plan it is necessary to adopt a flexible approach. This approach will help to cope with new problems and policies and to induct new technique in the management. Forest management is a complex task dealing with social, economical, cultural, legal, institutional, physical and ecological problems. Difficulties arise during implementation. It has been found that in order to fine tune the plan, it becomes necessary to revise and recast original strategy and objective. Therefore, it is imperative to make the plan flexible; so that any additional experience or knowledge gained in the process of executing the plan can effectively be incorporated in the management plan. In order to achieve this, the component of monitoring and evaluation process, in addition to other elements, should be built into the plan.

Planning Levels and Functions

Planning process begins with identification of determination of goal and objectives. Such an exercise depicts the purposes, premises, values and direction of planning. Once the goals and objectives are decided, the next stage of planning follows. Planning in an organization requires well-defined and delineated frameworks. Different organizations have different frameworks for planning. Generally the framework for planning in a forest department has three levels of management hierarchy. The planning functions of these three levels are as follows :

1. *Top level management (State level).* It determines strategic goal and formulates strategic plan for the forestry sector at the state level.
2. *Middle level management (Territorial Divisional and circle level).* It determines tactical goal and formulates tactical plan.
3. *Lower level management (Range level).* It determines operational goal and formulates operational plan.

The national and state forest policy determines the organizational goal and it sets stage for drawing up a strategic plan for the state. Now this strategic plan becomes the inputs for establishing tactical goals. Based on the goals, the tactical plan is formulated. Similarly, tactical goal determines operational goals. Based on these goals, the operational plan is drawn.

Types of Plans

Organization dealing with management defines goals at three different levels. Therefore, in order to achieve those avowed goals at different levels the organization requires formulation of plans at those three levels. Such plans are called as strategic, tactical and operational plans.

Strategic plan aims at achieving strategic goals. It illustrates the ways in which the organization will utilize its resources and conduct its activities in order to achieve the strategic goals. The top functionaries of the organization formulate strategic goals. Basically, it covers the long-term perspective planning.

Tactical plan aims at achieving tactical goals. It encompasses specific segments and aspects of the strategic plan. Suppose, the strategic plan comprises a component of rehabilitation of degraded forests. Then the tactical plan identifies it as one of the objectives of the plan. It aims at evolving suitable strategy to address it. Tactical plan, therefore, defines major problems already identified in the strategic plan and draws treatment to improve on the situation. Presently, in the departmental organization forest management plan (working plan) constitutes the tactical plan. Middle management functionaries prepare the tactical or working plan in the forest department. The middle level manager plays the role of tactical planner who is called as working plan officer in the department.

Operational plan aims at achieving operational goals. It supports the tactical plan. It comprises actions that are necessary for the achievement of operational goals. Lower management functionaries like forest range officer prepares such plans. Operational plan covers a short time frame and it serves as a guide for day-to-day operations. The harvest plan, plantation project, fire management plan, thinning plan etc. are some of the examples of operational plan.

Planning Process

Planning is the beginning of management process. Management as a whole is defined as the process of planning, organizing, leading and controlling the resources of an organization in the effective and efficient pursuit of specified organizational goals. This definition of management amply illustrates the critical relationship between planning process and other management process. Therefore, management of any resources begins with planning. Planning helps to identify what must be done and how it must be done. In other words, planning sets the stage and becomes central to other key management practices of the organization.

Goal Setting Process in Forest Management Planning

Goal setting is the beginning of the planning process. There are three basic management related issues in the goal setting process. These issues are critical to successful delineation of goal and objectives. The issues are illustrated below :

1. Dealing with multiple goal and objectives.
2. Barriers to goal and objective setting process.
3. Improving the goal setting process.

Forest organization has many goals and objectives to achieve. Most of the time these goals and objectives are not complimentary to each other. Quite often they are in conflict to each other. For example biodiversity conservation and timber production are two conflicting goals. Activities that aim at conserving bio-diversity and wildlife will be at the cost of timber production. Such conflict and contradiction appear on the way of determining goals and objectives.

However, such problems are resolved if the planner addresses himself to assessing the intensity and implication of such conflict and contradiction. After understanding the conflicts and constraints in its perspective, the next task is goal optimizing or balancing or reconciling of possible conflicts among different goals and objectives. The planner has to decide whether it is appropriate to undertake the fullest attainment of one goal or objective or a group of goals and objectives to the exclusion of others. In a forest organization, which is fully controlled by the government, the best choice is to pursue a course of action designed to attain a modest mix of all the goals and objectives.

The goal setting process in the forest organization suffers from two important infirmities. They are : (i) selection of unattainable goals and objectives; and (ii) overemphasis on either quantitative or qualitative goals or objectives. However, the goal setting process can be improved by adopting the following guidelines (Griffin, 1990).

- Planner should understand the purpose of goal.
- Goal should be properly stated :
  1. They should be specific.
  2. They should be concise.
  3. They should be time tested.
- Goals should be horizontally and vertically consistent.
- Planner must accept and be committed to the goals.

Working Plan Goals and Objectives

Forests in India are owned and managed by government. Therefore forest ownership is public in nature. Public ownership of forests enjoins upon government a responsibility to manage forests in such a way that it maximizes generation of public goods. Over the years, government has managed forest keeping the interests of the public at large. More specifically the forests in India are managed for production as well as protective purpose. In the past the production role took precedence over protection. However, at present, the role reversed. Forests are now managed with a primary objective to protect the entire ecosystem on which the survival of mankind depends.

Besides, with the introduction of New Forest Policy in the year 1988 a new era dawned in the forestry sector. Forest management focused on conservation of forest resources with the help of community participation. The place of production forestry and revenue yield has been relegated to a position of low priority. With New Forest Policy in force the instrumentality of management has changed. There is a shift from unilateral management to participatory form of management. Such changes in the paradigm of forest management necessitate proper determination of goals and objectives of management for a forest division. Forest management plan or forest working plan is an instrument by which the goals and objectives are delineated on the basis of policy environment and resource status.

Goals and objectives are central to the management plan. They constitute the framework for determining what action to take when they will be taken and the budget and personnel required to implement it. A management plan is a valuable tool for assessing the resource strength, identifying management needs, setting objectives and priorities and organising resources to achieve the goal. The management plan helps the manager to allocate and make best use of resources available at hand. The plan can also be used to assess the requirement of financial resources. A management plan can also serve as a communication tool to gain understanding and support from both the public and relevant government agencies. Such two-way communication helps in gaining the co-operation of local communities and the political goodwill. The planning process helps the management personal to comprehend the full spectrum of management needs. This ultimately provides a rational base for administration and implementation.

Objectives or goals are the ends towards which activities are aimed. In other words, they are the results to be achieved. They represent the direction towards which all management functions are aimed at. Strategies are general programmes of action and deployment of resources required to attain comprehensive objectives. Management is the means by which the objectives are achieved. If any of the avowed objectives are not redeemed, then there is a need to go for a review and evaluation. In such situation, there is either poor or wrong management inputs, or poor or wrong choice of objective, not achievable given the resource status and the existing ground realities.

Characteristics of Objective

Determination of objective constitutes the most important facet in a planning process. Before embarking on the task, it is imperative to understand the characteristics of objectives. Objectives have the following characteristics :

- The objectives should be clearly defined. They must be reduced into black and white. A clearly defined objective provides the clear direction for managerial efforts.
- The objectives must be realistic, *i.e.*, they must be within the reach of the organisation.
- The objective has to be written in the realm of possible achievement through strategies.
- Objective must be measurable.
- Objectives must have social sanction and acceptability.

Conservation necessitates long-term planning perspective. That means, the plan document should deal with two or three time scale or temporal entities. The short-term life of a plan, usually ten years, draws detailed scale of implementation. The medium term scale, covering two or three planning periods, requires different planning inputs. However, in the context of long-term perspective planning, the conservation goals and objectives are identified and determined and broad strategies are evolved to reach those objectives at a very long time scale.

Strategies are drawn to achieve the objectives. It is always imperative to evolve strategy compatible to ecological, social and political niceties of the said environment.

Management is an art of trade-off. Forest management throws a host of conflicts of objectives as well as interests. Therefore, management aims at resolving such conflicts and settling controversies. Generally, in order to achieve this objective, it is necessary to develop alternate resources and sustainable land practices for people depending on forests. Therefore, a management plan should not be written in isolation of other land-use, conservation and development policies.

National policies always set broad conservation goals. Legislation provides the legal basis for resolving conflicts that emerge during the course of management. Therefore, state policy and legislation set guidelines for evolving management practices and resolving conflicts. Such guidelines cover issues remained untouched in the central policy. State policy also takes cognisance of the resources of the state and proposes major strategy. The state policy deals with area like tourism, conservation, education, crop compensation, resources sharing, social security measures in protected areas etc.

Objectives of Management

Forest management at divisional level is the sum total of three distinct and separate activities. These activities are :

1. Translation of forest policy into a plan,
2. Execution and implementation of plan, and
3. Exercising administrative control over the plan.

Objectives occupy an important place in forest management process. They determine the shape of future events. Lack of objectives or failure to keep them clearly in the mind makes the task of management more difficult. By establishing goals and objectives the planner provides the spotlight over the courses along which management practices and activities are to be organized. In fact, an objective has an inherent power to stimulate action. Objectives help us to secure a clear vision and unity of purpose.

Management is the means by which objectives are achieved. If an objective is not reached, then there is either poor or wrong management inputs or, poor or wrong choice of objectives which is not reachable given the potential resources and the existing ground realities (Sawarkar,V.B.,1995).

Determination of objectives and their optimization provides a framework for evolving strategy for sustainable forest management. While determining the objectives of management for a forest division, emphasis should be given to clarity and distinctiveness. Besides, it should also be ensured that the objectives are in consonance with the established forest and environmental policy and socio-economic tradition of the locality. Deciding of objectives provides a basic framework for evolving comprehensive strategy for sustainable forest management. While determining the objectives of working plan, it is done at two stages. One is at the stage of beginning and second is at the stage of forming working circles. The objectives so determined in the outset are called general objectives of management. The objectives set in the second stage are called special or specific objectives of management. General or creed-type objectives symbolize social viewpoint of forest management. They have less significance; so far operational part of management is concerned. Special or specific objectives are the real objectives that concern the management very much. But specific or special objectives of management are always subordinate to the general objective of management of a working plan. (Table1.2). The general objectives of forest management are enumerated below :

1. Maintenance of environmental stability.
2. Preserving the natural forests along with biological diversity and genetic resources.
3. Checking soil erosion and conserving moisture.
4. Increasing forest area by afforestation.
5. Meeting the requirements of fuelwood, fodder, minor forest produce and small timber of rural and tribal population.
6. Increasing the productivity of forests to meet essential national needs.
7. Encouraging efficient utilization of forests produce and maximizing substitution of wood.
8. Managing forests with people's participation, for achieving those aforesaid objectives and to minimize pressure on existing forests.

Forestry is a long-term enterprise. It necessitates long-term planning. The instrument that epitomizes long-term planning is called Working Plan. It helps to develop supporting socio-economic and political environment necessary for promotion of sustainable resource management. Basically, this is a plan evolved for the sustainable management of forest resources of a division. For the sake of analysis and conceptualization, sustainable forest management connotes different things for different forests.

Forestry in India has an old tradition of initiating management through planning. This is how "the working plans or management plans" are in vogue in the forest departments since a long time. The practice of preparing and implementing such management plan generally covers all aspect of the forests. However, over the time, the practice of preparing separate management plan in the field of wildlife conservation has gained currency. A wildlife management plan is a product of planning process that sets out the values and objectives of management for a Protected Area (PA). The plan provides ways and means to achieve the avowed objectives within a fixed time frame.

Management plans have definite functions. They are as follows :

- It reveals the resource status.
- It reveals the trends of resource use.
- It lays bare the objectives of resource management.
- It expresses the problems and constraints.
- It provides an array of management options.
- It helps in evolving the optimal strategy.
- It helps in monitoring the resource use and accomplishment of objectivess.

The Need to Involve People

Management decisions are bound to affect different groups of people in different ways depending on what their interests are. If the management of natural resources is to succeed, it must be built on participation process, a mechanism to enlist support from the local communities.

Participation of people means involving people in natural resource management and understanding their problem and perceptions and developing strategies that aim at reducing the stress on people and pressure on forests. With reference to management, it is said that good forest management is synonymous with effective management of people's pressure on natural resources. It the pressures on forest resources are not reduced to the ecologically sustainable level and to the satisfaction of the social, economic and political aspirations of people, there would be no habitat left to manage. Legal measures and rigid protection are important but they generally lead to social confrontations. Therefore, there is a great need to tread cautiously while resolving conflicts. The art of management is to enhance management capability through people friendly, pressure dissipating strategies, which combine public education with actions that bring about mutually satisfying results.

Defining Working Plan

Management of any resource requires a management plan. Forest, being a important resource since time immemorial necessitated a management plan for generation of sustained timber yield. This management plan in forest parlance is called "Working Plan". A management plan is largely about resolving conflicts, choosing goals, objectives and making decisions. If conflicts cannot be resolved, objectives are chosen wrongly or out of context and if decisions are wrongly taken, the plan will be a failure (Sawarkar, V.B., 1995). Working Plan had been defined variously by different authors in different context and at different point of time. This is the reason why there is wide variation in their content and tone. However, few of such definitions are given below :

1. Working Plan is a written scheme of management aiming at continuity of policy, controlling the treatment of a forest. (Glossary)
2. Working Plan is the simplest possible statement of what is known about the working plan areas; its configuration, soil climate, vegetation, its possibilities; what has been done in the past, what should be done in future, how it should be done and what records should be kept.
3. A Working Plan sets fourth the purpose with which a forest should be managed so as to best meet the interests, and therefore the wishes of the owner; and indicates the means by which this purpose may be accomplished. In other words, it is a forest regulation prescribing the application of certain cultural rules and the execution of certain works, in order to produce a given result (D'Arcy, W.E.,1910).

Functions of Working Plan

Working Plan performs variegated functions. Few of such functions as enumerated by the Forestry Commission are :

- To ensure that National Forest Policy is made effective at all levels of forest management.

- To ensure the orderly and economic development of the forest within the framework of exiting policy and to prevent uncoordinated adhoc management decisions being taken without understanding their wider implication.
- To ensure that the aims of management are clearly defined and that the methods of achieving them are realistic and are the results of the careful study of all relevant factors.
- To ensure that all supervisory staffs concerned with the forest are aware of its intended future development and thereby achieve continuity of development when there are changes in staff.
- To bring together in a readily accessible form such information as necessary for the efficient management of the forests.

The aforesaid functions are primary in nature. Besides, there are other secondary functions of Working Plan too. Such functions are:

- To co-ordinate all the planning activities for the development and growth of the forests.
- To assess and compute status and potentialities of a forest.
- To provide a convenient repository for the data required in the day-to-day management of forests.
- To examine the constraints imposed upon the forests, the various possible courses of action and to work out which particular combination of possibilities most successfully achieves the management objectives.
- To convert objectives of management into achievement by means of annual plan of operations and allocation of budget, which relies on the needs, identified during the planning process.
- To provide a mechanism by which the higher management may exercise a fair degree of control over the type of operation undertaken.
- To provide an instrument to effectively monitor and evaluate the forestry operation.
- To induct modern technology and tools of analysis, monitoring and evaluation into the operation system.
- To understand the effects and impact of past management on the forest resources and the people living around forests.

Relevance of a Written Plan

Planning for forest management is a process of observation, appraisal and analysis with respect to resources, organizational strength, and problems/constraints. It also involves a process of charting a compatible course of management action. This process is learnt much better if the results of thinking and reasoning are written down in a simple but organized way. Apart from this, the fact that sustainable development of forestry, which needs a long term planning, calls for a systematic, logical scientific and analytical plan. Such plans are executed for a longer period of 10 to 20 years. Therefore, considered in the backdrop of those two factors, it becomes imperative to have a elaborate written plan for managing forests.

Changing Role of Working Plan

Concept of managing forests has changed with the passage of time. Since the inception, the primary objective of drawing a management plan was to maximize timber production from the forests. Therefore, working plan aimed at evolving a felling programme ; so that the entire accessible part of the forest would be taken up for harvesting wood in a definite period. In other words the main focus was to hammer out a harvesting plan. However over the years, in order to obtain sustained yield of timber on a long-term basis, regeneration strategy was amalgamated into the working plan. This led to the emergence of working plan of sort. During this period the methodology of computing the timber projection was basically very archaic and was based on thumb rules and cursory observation. Employing this methodology the working plan used to draw timber yield projection. This practice continued till the beginning of twentieth century. Afterwards, the practice of partial enumeration of trees and other improved methods to calculate timber yield began in the field of forest management. The exercise of computing timber yield projection on the basis of partial enumeration symbolized an improvement in technique and methodology. This method of estimation of timber continued till 1960 when the methodology of preparing detailed inventory of tress resources came into practice. Presently, this practice of conducting tree resource survey during management planning process is in vogue all over the country.

Over the years the role of forest changed radically. Timber

producing function gradually relegated to a place of low priority. Societal and protective objectives of forest management took precedence over other production and revenue earning objectives. Societal values of forests gradually changed with the passage of time. Considered in this backdrop, the necessity for adjusting plans to meet changing circumstances has assumed immense relevance and significance today than in the past. But what is disconcerting is the fact that today the situational framework where forest management works is more conditioned by a host of intractable problems and constraints. Therefore, there remains limited room for manoeuvring and consequently little choice as to the future course of action.

Components of a Working Plan

A working plan has to describe the situation, present the argument for and against different courses of action and prescribe for the future. Therefore, traditionally the plan document is organized into different parts accommodating separate aspects.

The first part comprises a number of chapters, each of dealing with a particular subject. This part is brief but deals in total relevance to the subjects. In other words, the chapters in this part should display a great deal of error-free data collection and analysis. The second part of the plan document is the place where various possibilities are discussed and important prescriptions are recorded with its accompanying analysis and justification. It is, therefore, the heart of the plan. The third part encompasses the entire appendix comprising data and maps prepared during the course of working plan process. Apart from these traditional parts, other reports like forest resource survey (inventory), forest soil resource inventory, joint forest management evaluation report and case studies on specific practices may be appended with as an additional part of the report.

Stages in Working Plan Formulation

The final Working Plan document is the product of meticulous planning process. In the process of formation of plan, there are many important and distinct stages, which assume significance (Figure 1.1). Such stages are elucidated below :

- ***Basic Information Report Stage*** **:** This is the initial stage of the Working Plan process. In this stage basic secondary information regarding forest division are collected.

- ***Preliminary Working Plan Report Stage*** **:** This is the second stage of Working Plan process. This stage again divided into two sub-stages. They are: First Preliminary Report Stage and Second Preliminary Report Stage.

Appointment of Working Plan Officer marks the beginning of this stage. Generally, this stage spans about 3 to 6 months. This initial phase culminates in formulation of First Preliminary (working plan) report. The report is then presented to a review committee constituting

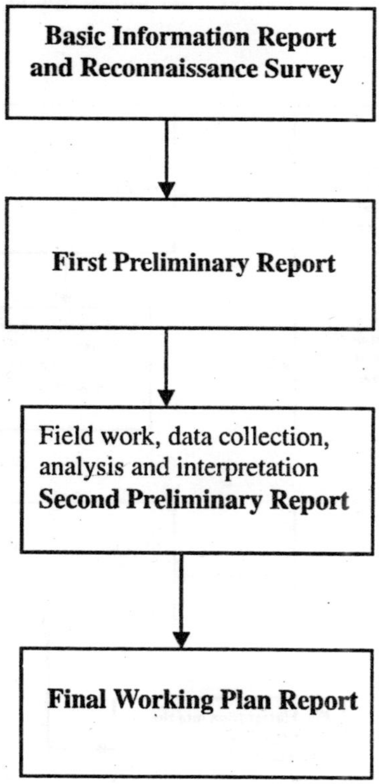

Fig. 1.1 : Stages of Working Plan Formulation Process.

senior officers of the state forest department and a representative officer of Ministry of Environment and Forests, Government of India.

A conclusive discussion and decision about the issues raised in the First Preliminary Report marks the beginning of the next phase of this stage i.e.; the Second Preliminary Report Stage. This phase is characterized by extensive data collection, analysis and interpretation. Apart from this, based on interpretation, the future strategy for sustainable forest management is drawn in the form of concrete proposals. Therefore, the Second Preliminary. Report encompasses analysis of social-political environment, resource status, problems/ constraints and future strategy for forest management. Under the strategy the process of working circle formation is undertaken (Figure 1.2). Submission of Second Preliminary Report to the review committee culminates this phase. This phase spans about 12 to 18 months.

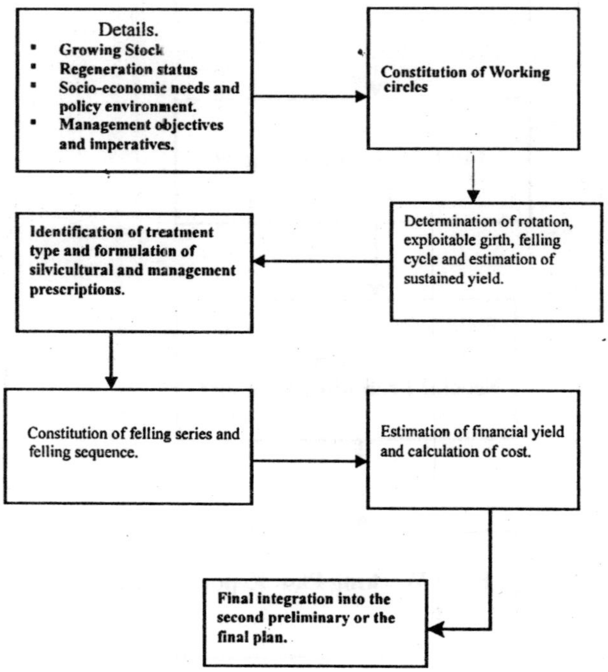

Fig. 1.2 : Process of Working Circle Formation

The Final Working Plan Report Stage

As the name indicates, this is the final stage of the working plan process. A conclusive discussion on the issues raised and strategy proposed in the Second Preliminary Report marks the beginning of this final stage of planning process. Generally, this stage spans about 6 to 8 months. Submission of final Working Plan Report marks the end of planning process.

Organization of Working Plan Areas

Forests are territorially classified into different categories for the purpose of description, administration, management and record. However, these sub-divisions are made primarily for the following purposes :

- For organizing forest operations.
- For controlling yield.
- For defining territorial responsibility.
- For providing, a permanent basis for locating forests and maintaining records.

The former two types are known as management and the latter two types are known as administrative classification. Both these categories, the management and administrative classification of forests assumes significance in the context of working plan (Table 1.1). However, the management types of classification assume more importance in the course of working plan formulation process. Sub-divisions of forest from the management point of view are elucidated below :

1. Working Circle or Management Circle.
2. Felling or Treatment Series.
3. Compartment.
4. Periodic Block.
5. Coupes.

Table 1.1 : Sub-division of Forests

Administrative sub-division	*Management*	
	Sub-division	*Function*
• Working plan areas • Range • Circle • Beat • Compartment	Working circles	Organization of work
	• Felling series • Treatment Series • Coupes	• Silviculture • Yield regulation

Working Circles

The forests of a division constitute a unit. The area of the unit is normally large. The unit contains heterogeneity in terms of site condition and crop composition. Therefore, in order to improve the growing stock, different silvicultural treatments are required for different parts of the plan area. As a result different set of prescription or working rules are required to be drawn up for different parts of the forests. It is because of this reason that the forest is divided into different category of areas on the basis of their required management practices. Such categories of forest areas are called working circles.

A working circle is an area of similar or more or less uniform crop, which may form the whole, or part of a working plan area. In other word, it is one of the sub-divisions of the working plan area. Generally, reasonably similar crops are put together to form a working circle where a common treatment can be applied. In the text books of forestry, working circle has been defined as, 'a forest area (forming the whole or part of the working plan area) organized with a particular object and under one silvicultural system and one set of working plan prescriptions. In certain circumstances the working circle may overlap'. As for example, the bamboo forest generally constitutes an overlapping bamboo working circle. However, over the years, with the decline in forest crop, the concepts of working circle formation has changed. Now, generally, working circle constitutes forest areas which generally requires a common treatment. For example, all forest areas that need rehabilitation with varying intensities are grouped in one or two rehabilitation circles. Likewise all forest areas, which need artificial regeneration, are grouped in one working circle popularly called as plantation working circle.

Felling Series

Concentrated felling in one forest area in a year shall create labour, transportation, social, marketing, staff, and other operational problems. On the contrary distributed felling emits continuous benefits in terms of employment, to rural folk and precludes prohibition of grazing in the forest area on a continuous time period. Therefore, it is desirable to distribute the felling all over the working circle areas. Such a scheme goes in consonance with social, political, economic, and operational requirement. In order to achieve this objective the working circle areas are divided into different felling series areas.

Felling Series (FS) is a forest area forming the whole or part of a working circle and delimited so as: (1) to distribute felling and regeneration to suit local condition, and (2) to maintain or create a normal distribution of age classes. The yield is calculated separately for each felling series, which should have an independent representation of age classes (BCFT. Modif). Each felling series is deemed as a self-contained unit of management with a separate calculation of yield and a separate series of silvicultural operations. Division of working circle areas into such felling series facilitates effective control and distribution of work into different ranges.

When the forest area does not have enough growing stock for felling, then in such cases the working circle area is divided into different treatment series. Treatment series are group of compartments, which requires analogous treatment in terms of rehabilitation and regeneration. Such treatment series generally provides a framework for systematic and planned treatment of degraded forest areas.

Compartment

Compartment is a territorial unit of a forest permanently defined for purposes of administration, description and record. It is also the unit of management area. Generally, a compartment is given one set of treatment. This is sf a temporary nature, differentiated for special description and treatment (Glossary).

Coupe

Coupe is the unit area in a felling or treatment series taken up for felling or treatment every year. A group of compartments or a compartment or a part of compartment may form a coupe. While determining area of coupe it is necessary to decide the quantum of work the treatment requires and its operational feasibility. In a working circle the number of annual coupes would be equal to number of felling or treatment series.

Periodic Block

In the past uniform silvicultural system was in vogue in many parts of the country. Under this system, there was a practice of dividing the forest area into periodic blocks, which symbolized different age classes. Periodic block has been defined as the part or parts of forest set aside to be regenerated, or otherwise treated, during a specified period. The regeneration block is called "floating" or "single" when it is the only periodic block allocated at each working plan revision. When all periodic blocks are allocated and retain their territorial identity at working plan revision, they are termed "fixed" or "permanent" (Glossary).

Notwithstanding the territorial units in preparing a management plan, there are many important management terms, which are extensively used to define management practices. Few of such terms are illustrated below :

Felling Cycle

In a selection forest the usual practices is to divide the area of the working circle into a number of coupes. Such coupes are worked at an interval of definite period of years known as "Felling Cycle". Felling cycle has been defined as the time that elapses between successive main fellings on the same area (Glossary). It may vary from 10 to 30 years or more. Its determination depends on the silvicultural needs of the crop as well as the objectives of the management. If the objective is to produce timber of large size, then a longer felling cycle is opted. A smaller felling cycle produces small timber.

Exploitable Diameter or Girth

In selection forests, few trees, which have matured, need to be removed; so that more light will penetrate into the floor of forests. This will help promote natural regeneration. This apart, the existing young and middle-aged crop shall get more space for diametric growth. Therefore, in a selection forest a specific girth or diameter of the tree, which indicates maturity, at least from the point of management, is considered suitable for exploitation. The specific girth or diameter is determined by rotation of maximum volume production method (Bebarta, K.C. 1999). This exercise of determination aims at optimizing the yield of timber crop. Therefore, in timber management practice the exploitable diameter or girth symbolizes rotation of maximum volume yield.

Formation of Working Circles

The strategy for forest resource management begins with the formation of working circles and miscellaneous regulation. Therefore, formation of working circle is an important step in the planning process. By tradition, Indian forests are managed by a system of working circles. Formation of working circles involves many steps (Figure 1.2). These steps are as follows :

1. *Determination of working circles :* Working circles are determined on the basis of ground requirement and objectives decided.
2. *Naming of working circles :* Working circles are generally named on the basis of regeneration method adopted. However, over the years, nomenclature process has adopted a functional approach. Both the practices are still in vogue. Selection working circle aims at regenerating the crop on the basis of selection system of regeneration. Likewise, rehabilitation working circle aims at rehabilitating degraded forest by artificial regeneration and subsidiary silvicultural operation. A plantation working circle assumes its name because it aims to improve the stocking by plantation.
3. *Allotment of compartments into different working circles :* This is another important step in the working plan process. After stock mapping and assessing strength and weakness of every compartment is completed, the next logical step is their allotment into different working circles.
4. *Determination of rotation and felling cycle :* Rotation and exploitable diameter assume significance in case of timber

management working circles. Different kinds of rotations are decided on the basis of objectives chosen (Table 1.2).

Table 1.2 : Objectives and Rotation in Different Working Circles

S. No.	*Working circle*	*Principal objectives*	*Rotation*
1	Selection-cum-Improvement	(i) Conservation	Physical rotation in conservation and erosion prone areas
		(ii) Timber production	Rotation of maximum volume production
2	Coppice with Reserve (CWR)	(i) Conservation	Physical rotation in conservation and erosion prone areas
		(ii) Timber production	Rotation of maximum volume production
3	Rehabilitation of degraded forests	Conservation	Physical rotation
4	Biodiversity Working Circle	Conservation	Physical rotation

Felling cycles are decided on the basis of crop growth and special objectives of the management. While choosing felling cycle, it should be borne in mind that the period should synchronize with the regenerative capacity of the crop. A short felling cycle may remove the crop faster than it is substituted.

5. *Determination of exploitable girth/diameter and selection of species for exploitation :* This is another important step in the working circle formation process. In the beginning, it is necessary to select the species, which are required to be exploited in the working circle areas. This is decided on the basis of crop growth, crop composition and special objectives of management.
6. *Estimation of annual increment or yield :* This is the next step in the working circle process. The probable sustained yield with respect to each working circle or felling series is calculated on the basis of yield estimation formulas.
7. *Identification of treatment type and silvicultural prescription :* As we know, forest management pursues multiple objectives. A single unit of forest can have more than one objective. Each compartment in the forest contains different

opportunities and problems. Therefore, each such unit, be it a compartment or coupe, is divided into different treatment areas. Such treatment areas are geographically demarcated and prescription for each such treatment type is formulated. Different treatment types used in management practices are presented in Table 1.3.

Table 1.3 : Treatment Types in Coupe or Compartment

Treatment type	*Description*	*Objectives*
Type-A	• Erosion areas • River banks • Sloppy areas. • Conservation areas	Protection
Type-B	Dense areas	Production/bio-diversity conservation
Type-C	Congested pole crop areas.	Improvement felling and thinning
Type-D	Degraded areas	Rehabilitation

8. *Constitution of felling series and felling sequence of annual coupes :* After having allotted compartments into different working circles, the next step is to divide the working circle areas into felling series. Felling series are formed on the basis of : (1) operational feasibility for carrying out silvicultural works, and (2) yield prospects. Once the felling series areas are delineated, it is time to decide on felling sequence. Determination of felling sequence depends upon crop condition, yield potentiality and time when the coupe was harvested in the last plan. After this process, all the annual coupes are again plannimetered and management maps are prepared.

PART II

Study Area

Rajnandgaon district constitutes the study area. It is one of the revenue districts in the state of Chattisgarh. It is situated in western part of the state. Till the year 2000, the revenue district had only one territorial division, *i.e.* Rajnandgaon forest division. But after the formation of Chattisgarh in the year 2000 Rajnandgaon forest division was

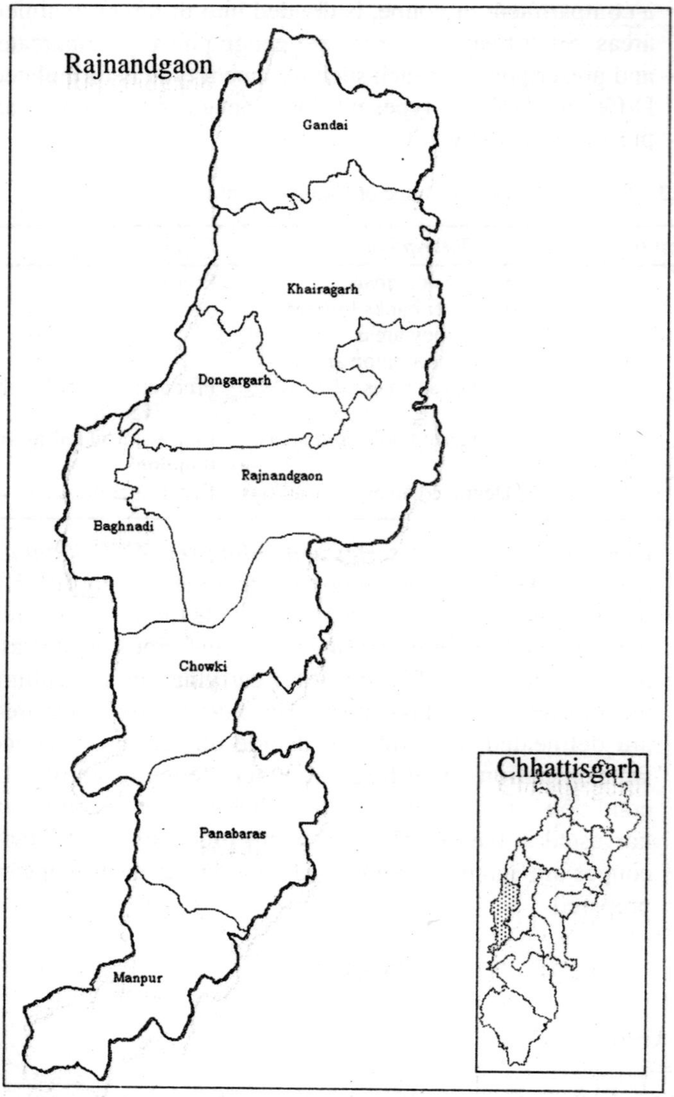

Fig. 1.3 : Map of Rajnandgaon District

bifurcated into two forest divisions: Rajnandgaon and Khairagarh (Fig. 1.3). It constitutes a landscape from the point of natural traits such as landform, soil characteristics, watershed attributes and forest resources. A detailed account of the study area is presented in chapter two.

Methodology of Forest Management Planning

The study area encompassing the resources of districts constitutes a landscape. It has all the attributes of a landscape. It has wide diversity of topography, rock, soil and forest resources. The author prepared the working plan for the study area for a period of ten years, *i.e.* 1999-2009. A well established methodology which had been in vogue in the department since twenty years was adopted to gather and process the field data. The methodology consisted of the following steps.

1. Delineation of study area.
2. Designing working plan methodology and processes (1st Preliminary WP).
3. Conduct of field works.
4. Data entry, and data analysis with help of specialized application software.
5. Constitution of working circles and formulation of prescriptions (2nd Preliminary WP).
6. Constitution of annual coupes (Preparation of Management map sheets).
7. Preparation of final Working Plan report.

The study area has been described in separate chapter. It deals with an in-depth analysis of demographic, topographic and forest resource data. This provides the context for evolving management options. Fieldworks have been conducted and necesary data were collected. These data were fed into specially prepared application software for analysis and reporting. Based on the data regarding forest resources and stock mapping, working circles were constituted. The broad principles of forest management on which the prescriptions were evolved are presented in chapter four. Constitution of working circles and its basis has been discussed in chapter five. Scope for participatory forest management and agro-forestry in the study area has been discussed in the last two chapters of the book.

Conduct of Fieldwork

In order to evolve best options for silvicultural management, it is imperative not only to analyze the impact of past management on forests but also to assess the present state of forest resources. This requires sufficient data for analysis and assessment. Therefore, fieldworks were conducted for collecting data.

The following fieldworks were undertaken during the working plan revision process :

1. Reconnaissance survey
2. Compartment history writing
3. Stock mapping
4. Socio-economic survey
5. Survey and assessment of natural resource :
   - Sampling for bio-resource survey,
   - Regeneration survey,
   - Plantation survey,
   - Forest soil survey,
   - Stem and stump analysis.

1. *Reconnaissance Survey :* The author as a working plan officer undertook a reconnaissance survey of the study area. Before undertaking the job, the author inspected and examined the plan area, forest blocks, and compartments. The past management was studied and reconnaissance survey with respect to attributes such as crop condition, stratification, regeneration status, stem distribution, forest density and growth potentialities, was carried out.

2. *Compartment History writing* : Working plan revision requires that description of compartment history should incorporate all the important changes that took place during the last plan period. Description of features in compartment is made in a concise manner. It conveys an adequate idea of the composition, quality, age, density and general characteristics of the growing stock. The notes prepared by the range officer during their visit to the forests formed the main basis of compartment history and description. This again constituted the basis on which forest management for that compartment was based. Notes were taken as per the compartment history formats approved by State Government.

3. *Stock Mapping* : Stock mapping has been done accurately

traversing the compartment. This process helped to cover the entire area and to assess the site quality, species composition, crop density, crop age, plantation, etc. All locations where any change has taken place were depicted/delineated on the stock map. Regeneration status of main species has also been carefully observed and recorded in the prescribed format and also in the trace map. Basically stock maps were prepared by range officer and checked by Working Plan Officer.

4. *Survey and Assessment of Natural Resources* : Assessment of natural resources is an essential and integral part of working plan exercise. On the basis of analysis, past performance is evaluated and future management is proposed. In case of forest crops, it is required to identify multidimensional populations with various parameters and attributes. Forest resource data gathered during the working plan exercise are :

- Tree species stand or forest type through enumera-tion of growing stock.
- Crop composition: on the basis of age, density, extent and its distribution through stock mapping and enumeration of growing stock.
- Statistics of growth and yield (through stump and stem analysis) from primary and secondary data.
- Status and spatial distribution of regeneration through enumeration.
- Status of plantation by plantation survey.
- Forest soil type through soil resource survey or secondary data.
- Common non-timber forest produce.

Forest Resource Survey (Inventory)

Forest resource survey is a partial enumeration of tree resources with the help of sampling. It was done through stratified random sampling. The whole of planned area was divided into three strata, namely : Stratum-I, Stratum-II and Stratum-III. The stratification was based on management parameters. The first stratum consisted compartments which were allotted to Conversion, Selection-cum-Improvement (SCI) and Plantation working circles of the past plan. The second stratum consisted of compartments allotted to Improvement and Coppice With Reserve (CWR) working circles of the past plan. The third stratum consisted of compartments of the rest of the working circles of the past plan. In order to find out the required number of sample plot, a pilot survey was conducted by randomly selecting 50 plots having size of 0.1 hectare each. Based on the analysis of volume data gathered from these pilot survey plots distributed in the whole of the plan or study area, the minimum number of sample plots required for attaining 95 per cent probability with 5 per cent confidence limit under different strata has been determined (Table 1.4). Sample plots were laid at an interval of one minute fifteen seconds each on a topo-sheet of 1: 50,000 scale. The rectangular sample plot size was 0.16 hectare (Fig. 1.4). The sampling intensity remained at 0.16 per cent of the total forest area. Enumeration was done in 10 cm girth class interval, beginning from 21 cm girth at breast height and data were recorded in six different formats.

Table 1.4 : Number of sample plots in different strata

S.No.	*Stratum*	*Number of sample plots*
1.	Stratum-I	181
2.	Stratum-II	125
3.	Stratum-III	37
	Total	343

Regeneration Survey

Regeneration survey was combined with tree enumeration exercise. While carrying out enumeration, status of regeneration of main species at five square sub-plots having size of 2×2 metres falling inside the main sample plot was recorded in categories such as established, unestablished and recruit.

Plantation Survey

Valuable forest resources were created under various schemes of plantation. An up-to-date and reliable knowledge of these man-made forest plantations in respect of parameters such as species, growth and harvesting is necessary for evolving corrective measures. In general, a record of plantation is maintained in plantation journals and information regarding success per cent, mortality and growth recorded therein. Survey of plantation to collect primary and secondary

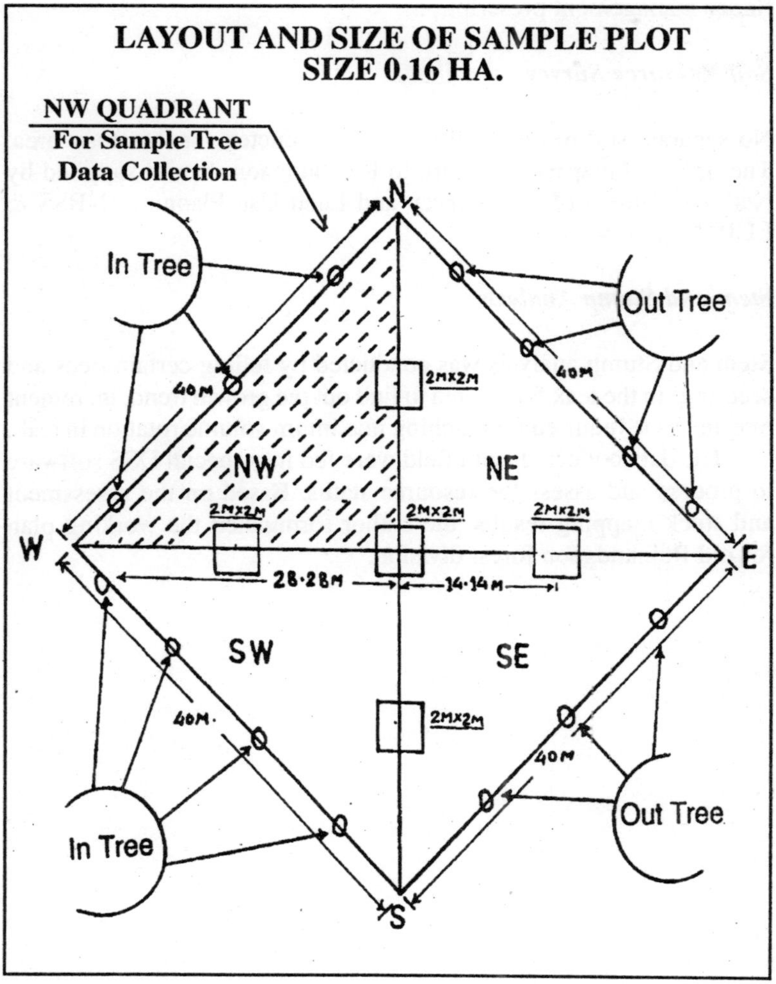

Fig. 1.4 : Layout of a Sample Plot

was carried out with partial enumeration to know the macro position of plantation in the study area. This knowledge helped in drawing future management prescriptions.

Soil Resource Survey

No separate soil resource survey was conducted for the study area. The data and maps with regard to Rajnandgaon district supplied by National Bureau of Soil Survey and Land Use Planning (NBSS & LUP) Nagpur were used.

Stem and Stump Analysis

Stem and Stump analysis was conducted by felling certain trees and seedlings in the teak forest area to find out the growth trend, increment rate and time required for reaching maximum volume rotation in teak.

The data collected from field were fed into special DSS software to process and assess the resource status. Based on the assessment and stock mapping results, the author formulated the working plan for old Rajnandgaon forest division.

2

Rajnandgaon : The Study Area and its Forest Resources

Management of any resources, including forests and wildlife, begins with planning. Planning helps in identifying what must be done and how it must be done. In other words, planning sets the stages and takes control of other key management practices of the organization. It comprises several closely related steps or activities. One among is the assessment of strength and opportunities of the area for which planning is undertaken. Understanding the history, climate, soil, demographic dynamics, natural resources and social customs of the study area is the beginning of forest management planning process. Therefore, this chapter delves in definition of landscape, history, location, land use pattern, factors of locality, demographic dynamics, social fabric, occupational patterns, disturbance regime, rights and concessions and forest resource assessment of the study area. This study area has been construed as a landscape.

Rajnandgaon district constitutes the study area. It is one of the revenue districts in the state. Till the year 2000, it used to be one forest division. But after the formation of Chhattisgarh in the year 2000 the area falling in the revenue district was bifurcated into two forest divisions: Rajnandgaon and Khairagarh. It constitutes a landscape from the point of natural traits such as landform, soil characteristics, watershed attributes and forest resources.

Defining Landscape

Landscape area is an integrated land unit bounded by natural boundaries such as river and watershed areas. Management of forests at landscape levels requires co-ordination of the spatial and temporal consideration of the silvicultural regimes for all stand included in the landscape. This co-ordination ensures that the proposed regimes are physically feasible within the spatial consideration in such a fashion that maximizes the process of commodity flows, distribution of stand structures, and delivery of social, cultural, economic and environmental values across the landscapes physically as well as temporally.

The study area qualifies to become a landscape. It contains all the characteristics of landscape. The spatial structure of study area, including biological composition, its physical environment and anthropogenic or social pattern indicates the repeating patterns associated with different land uses. The relative proportion of forests, agriculture, and urban land use makes the landscape unique. More to it, from the point of ecology, the landscape presents a mosaic of varying and interacting ecosystems, be it forests or agriculture or urban, distinguishable from one another on account of geology, soils, climate and human influences. In other words, the landscape contains a heterogeneous composition of multiple land units.

The study area is made up of watersheds, network of small ecosystems, forest and agricultural and group of habitations. All these features along with their interactive dynamics form distinctive landscape characteristics. The traits, patterns, and structure of the study area, including its biological composition, its physical environment, and its anthropogenic or social patterns indicating a dynamically juxtaposed ecosystems shall be discussed in the ensuing chapters.

History and Location of Study Area

The district Rajnandgaon came into existence on 26th January, 1973, by way of division of old Durg district. The Rajnandgaon state in the past was ruled by Somvanshis, Kalchuris and Marathas. The palaces in the town of Rajnandgaon reveals its own tale of the rulers, their society and culture and the splendid tradition during that time. The contribution of Gajanan Madhav Muktibodh, Padamlal Punnalal Bakshi and Baldeo Prasad Misra in the field of Hindi literature has a special mention. The district Kawardha was later bifurcated from the district on 1st July, 1998.

The district Rajnandgaon is situated in central part of Chattisgarh. The district headquarter Rajnandgaon is on the Bombay-Howrah line of South-Eastern Railways. The national highway 6 (Great Eastern Road) passes through the town of Rajnandgaon.

Situation

The study area takes into account all the Reserved, Protected and part of the Orange Area (Unclassified forests) except the forest area already transferred to Forest Development Corporation. It lies between 20° 48′ and 21° 48′ North Latitude and 80° 2′ and 81° 4′ East Longitude. It consists of two territorial forest divisions *viz*, Rajnandgaon and Khairagarh. Both the divisions consist of eight forest ranges such as Gandai, Khairagarh, Rajnandgaon, Dongargarh, Baghnadi, Chowki, Panabaras and Manpur.

The principal river of the district is Sheonath, a tributary of Mahanadi. The river originates in the forest areas along the border of Chowki block and flows in the north-eastern direction. The main tributaries of Sheonath are Kharkhara, Sonbarsa, Amner, Surhi, Karra, Murkati, Sankari, Fonk and Hanf.

Land-Use

The pattern of land use at any particular time is determined by the physical, economic and institutional framework taken together. In other words, the existing land-use pattern in the districts has been evolved as a result of the action and interaction of various factors, such as the physical characteristics of land, the institutional framework, the structure of other resources (capital, labour etc.) available and the location of the district in relation to other aspects of economic development. The present pattern can, therefore, be considered to be in consonance with other main characteristics of the economy of the state or region.

The total geographical area of the district is 9,208 sq. km. It has been classified into five broad categories known as five fold land utilization classification. The land use statistics of the district are presented in the Table 2.1 and Fig. 2.1.

Table 2.1 : Land-use Classification of District Rajnandgaon

Sl. No	*Item*	*Area in sq. km.*
1.	Total Geographical Area	9207.57
2.	Forest Area	2931.02
3.	Area not available for cultivation	2024.86
4.	Fellow land	139.85
5.	Waste land	460.5
6.	Net Area Sown	3651.34

Source : District Statistical Book, 1997. Rajnandgaon.

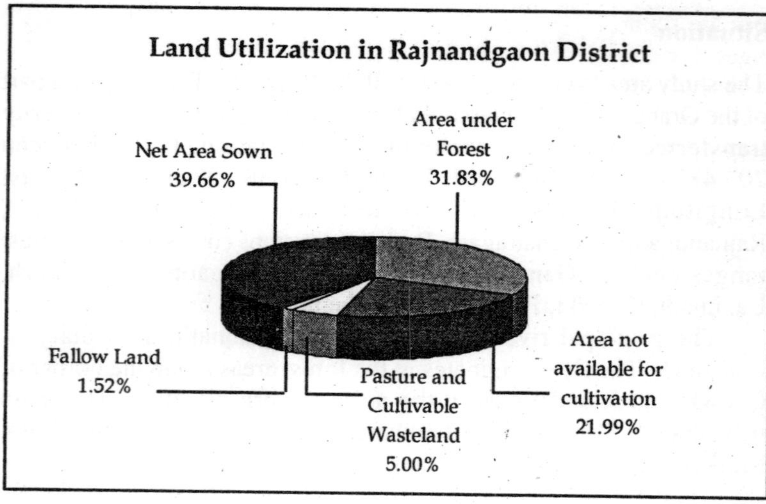

Fig. 2.1 : Chart Showing Land Utilization in Rajnandgaon District

Factors of Locality

The flow of existing benefits or potential benefits derived from a forest are determined by the conditions prevailing in the forest. An understanding of forests and factors locality constitutes an essential requirement for forest management. The type of forests occurring particularly on any site is very largely dependent on such factors as general climate, geology, and topography, which can hardly be altered by the forester, but it is also related to soil and biological factors which can be significantly influenced by the treatment he applies (Champion, G.H. *et al.* 1968). Factors of locality influencing the Study area are presented below :

Configuration of the Ground

Different types of topographical features are found in the study area. Satpura hill ranges lie in the northwest part of the district. Such hill ranges consist of steep hills, escarpments and spurs. Basically, they consists of granite/gneissic landscapes and basaltic landscapes. The Dandakaranya plateau extends upto south of the division. Majority of the study area belongs to category of hilly and undulating terrain. There is also plain area found in the eastern part of the division. Plain areas are found in the southern part of the division too. Majority of ranges like Gandai, Khairagarh, Dongargarh are hilly and the rest of the ranges comprises a combination of hilly and plain areas.

Geology, Rock and Soil

The rock formation of the area can broadly be divided into three groups: (i) older metamorphosis (unclassified), ortho, paragneises and scists, (ii) Dongargarh Super Group, (iii) Chhattisgarh Group and Deccan Trap Soil is generally deep along slopes and foothills. Major portion of the study area contains loamy, clayey and sandy loam soils.

Climate

The climate of the district is sub-tropical with three distinctly defined seasons (Table 2.2). The tract experiences rains from south-west monsoon from mid June to mid-September. The average rainfall of the study area is 1414 mm and number of rainy days are 65. The mean daily temperature is 25.5°C. The highest temperature ever recorded in the area is 46.7° C in the year 1989. The overall climatic conditions of the study area are suitable for fair growth of the plants.

Table 2.2 : Season and its Period of Spread

S.No.	*Name of Season*	*Period*
1.	Rainy season	Mid-June to September end
2.	Winter season	October to mid-March
3.	Summer season	Mid-March to mid-June

Temperature, rainfall and humidity of the study area were analyzed by using 'Liper' formula developed by Mr. A.A. Liper in 1951. The inferences drawn from the analysis are presented in the Box 2.1.

Box 2.1 : Range of Temperature in Rajnandgaon District

- The average annual temperature of the study area is 25.5° C.
- The average annual maximum temperature is 30.5°C and this temperature ranges from 26.8°C in January to 38.6°C in the month of May.
- The temperature in the winter season *i.e.* from November to February ranges from 25.7°C to 29.6°C and the average minimum temperature is 13.8°C to 18.3°C.
- In the rainy season, the average maximum temperature ranges from 27.7°C to 29.7°C and the average minimum temperature ranges from 22.7°C to 23.6°C.
- The coldest months are December and January.

Prospect of Plant Growth

Temperature and precipitation play a major role in plant growth. When the temperature rises above 48°C, it starts to leave an adverse impact on the plant growth. But a major part of the study area remains below this mark. Therefore, there is no adverse impact of simmering heat on the forests. Likewise, when the minimum temperature in the winter dips down below 6°C, than also it leaves an adverse impact on the plant. But the district does not experience this extreme range of temperature. Nevertheless, during the period from October to November and February to March growth is not conducive to plant. It is because the average minimum temperature during this period ranges around 15° C and does not exceed this point.

Rainfall pattern in the study area helps the plant growth. Heavy rainfall takes place in the month of July and August. Both the months experiences more than 50 per cent of the total rainfall. In contrast to this, the rainy days in these two months are less than 50 per cent of the total rainy days of the year. The intensity of rainfall in these two months is so heavy that it causes heavy run-off in the hilly part of the forests.

Geographical Area and People

The geographical area of the district Rajnandgaon is 9207.570 sq.km. The recorded forest area (on the basis of land revenue record) is 2931 sq km. It means the forests constitute about 32 per cent of the geographical area of the district.

The population of the district (undivided) was 10.9 lakh in the year 1991. The population density was 135. The per capita forest is 0.27 hectare. Literacy percentage of the district is 44 per cent. The population growth rate in the year 1991 was 22.43 and it is 17.7 in the last census 2001 (Table 2.3).

Population Dynamics

Population growth has a definite impact on sustainable utilization of natural resources. It is, therefore, pertinent to analyze the population growth pattern in the study area. The population growth of the district in the period 1981-91 was 22.43 and in the period 1991-2001 it is

Table 2.3 : Glimpses of Population Statistics of the Rajnandgaon District

Region	*Geographical area (sq. km.)*	*Population in 1991 & 2001 census*		*Density in 1991 & 2001 census*		*Population growth rate (%)*	
		1991	2001	1991	2001	1981-91	1991-21
Chhattisgarh	136030	17614,958	20,795,956	130	154	25.73	18.06
Rajnandgaon district	9208	1089,047	1,281,811	135	159	22.43	17.70

Source : Census of India 1991 and 2001.

17.70. It means there is a sharp decline in the population growth rate. But, the decline is not uniform in all the six revenue tahsils of the district. The decadal growth of population with respect to different tahsils is presented in Table 2.4.

Table 2.4 : Decadal Growth Rate in different Tahsil : 1991 Census

Sl. No.	*Tahsil*	*Decadal Growth rate in %*	*Name of Forest Range*
1.	Chui Khadan	23.79	Gandai and Khairagarh
2.	Khairagarh	21.62	Khairagarh
3.	Dongargarh	21.85	Dongargarh
4.	Rajnandgaon	24.19	Rajnandgaon and Baghnadi
5.	Ambagarh Chowki	17.47	Chowki
6.	Mohla	24.96	Panabaras and Manpur

Source : Census of India, 1991.

The population growth rate in Mohla tahasil which covers forest area of Panabaras and Manpur is the highest. Next to it, comes the Rajnandgaon and than comes Chui Khadan which covers area of Gandai and Khairagarh forest ranges. The growth rate is lowest in the Chowki tahasil. The population growth rate in the period 1991-2001 is 17.47. Growth rate in the rural area of the district is 18.50 and in the urban area it is 14.2. Though the growth rate has declined subsequently, but it requires further drop in rural sector because majority of population lives in rural areas which are in proximity to forests.

Population Density

Population density is a distinct indicator of anthropogenic pressures on the natural resources. In the year 1901, the population density was only 53 and it increased more than two folds in a period of about 90 years and reached 135 in the year 1991. According to 2001 census, the population density of the district is 159. It means, the population density of the district is increasing at 1.8 per cent annually. An account of change in population density in a period of 90 years is presented in Table 2.5 and Fig 2.2.

Table 2.5 : Change in Population Density of the District in the 20th Century

Sl. No.	*Year of census*	*Population density*	*Decadal growth in %*
1.	1901	53	–
2.	1911	63	19
3.	1921	57	–
4.	1931	64	12
5.	1941	72	12.5
6.	1951	76	5.5
7.	1961	96	26
8.	1971	89	–
9.	1981	105	18
10.	1991	135	22.43
11.	2001	159	17.7

Population Distribution

The distribution of population presents a heterogeneous picture. The population density shows much variation at the micro level. The tract carrying forest range such as Manpur has a density that is lower than the district average. It has the lowest density of 81 persons per sq. km. Dongargaon and Rajnandgaon tahsils are densely populated. The density is found to be the highest in both the tahasils. The population density in Dongargaon is highest at 212 persons per sq. km. (Table 2.6).

Sex Ratio

It is not known whether there is any relationship between sex ratio and forest development, but there is a definite relation between sex ratio and socio-economic development. The sex ratio in Rajnandgaon district shows an interesting gender scenario that is favourable to

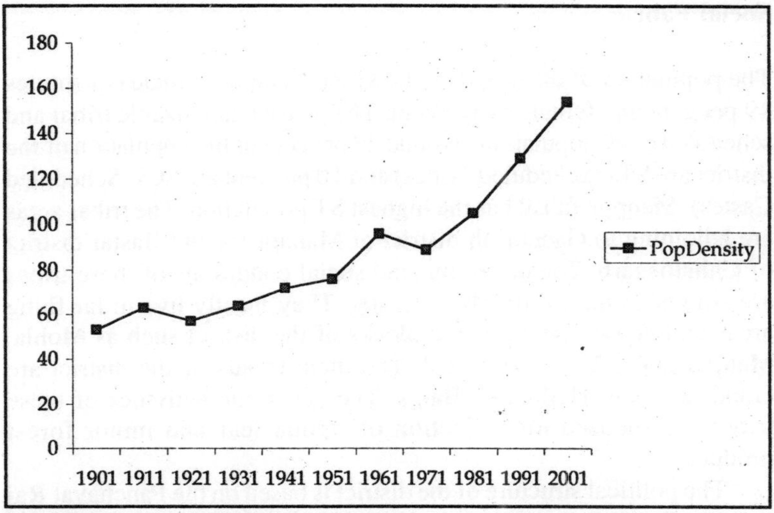

Fig. 2.2 : Population Density in Different Years.

women. In all tahsils except Rajnandgaon, the Female–Male ratio is very high. In Chowki tahsil, the ratio is highest. This indicates that the society has very favourable disposition towards the girl child. Possibly, because of this trend, the incidence of child marriage is also high in the district. According to 2001 census, the sex ratio in the district is 1024 (Table 2.6). It was 1016 in the 1991 census.

Table 2.6 : Sex Ratio (Number of Females per 1000 Males) in 2001

S.No.	*Tahsil*	*Population*	*Sex Ratio*	*Population Density*
1.	Chui Khadan	1,39,447	1023	127
2.	Khairagarh	1,56,283	1024	92
3.	Dongargarh	1,74,262	1025	105
4.	Rajnandgaon	4,53,591	1013	172
5.	Dongargaon	1,10,730	1032	212
6.	Ambagarh Chowki	94,365	1069	141
7.	Mohla	75,968	1044	81
8.	Manpur	77,165	1006	81
	Total	12,81,811	1024	135

Note : Population density has been taken from 1991 census figures and rest of the figures were taken from 2001 census.

Social Fabric

The population of the district is 12,81,811. Out of it, male constitutes 49 per cent and female 51 per cent. The district has sizable tribal and scheduled caste population. Around 27 per cent of the population of the district are STs (Scheduled Tribes) and 10 per cent are SCs (Scheduled Castes). Manpur Tahsil has the highest ST population. The tribal areas are adjoining to Gadchiroli district of Maharastra and Bastar district of Chhattisgarh. The economic and social conditions of these tribes are similar to that of tribals of Bastar. They mostly live in far-flung areas and dense forests. Three blocks of the district such as Mohla, Manpur and Chowki are tribal. The main tribals of the district are Gond, Kanwar, Halba and Baiga. The economic activities of these tribes are focused on collection of tendu leaf and minor forest produces.

The political structure of the district is based on the Panchayat Raj system of self-governance, but at the village level, leadership patterns mirror old traditional social systems. Each village has a Mukhia– a leader by social acceptance, other than the Panch or Sarpanch under the Panchayati Raj System. Any conflict among the villagers as well as any issue regarding the general development of the villages is resolved, and regulated under the leadership of Mukhia or a group of Panch.

The overall literacy rate of the rural areas of Rajnandgaon district is 48.77 per cent, and it is now 77.58 per cent as against the national literacy rate of 52.1per cent and state average of 65.18 per cent (Census 2001) (Table 2.7 and Fig. 2.3). Genderwise analysis, as shown in Table 2.7, reveals that 77.86 per cent of the rural male population of Rajnandgaon is literate compared to 52.40 per cent of rural women population. In contrast, the average male and female literacy rates in Chhattisgarh are 77.86 per cent and 52.40 per cent respectively.

Occupational Patterns

Occupational patterns and work participation rate in the district are note worthy. It shows a dominance of the primary sector in the economy of the area. Nearly 83 per cent of the total population of the district is dependent on the primary sector for livelihood, whereas the secondary and territory sectors account for 11 per cent and 6 per cent of the share respectively. In the case of agricultural and allied sector,

Table 2.7 : Population, and Literacy Percentage in different Tahsils

S.No.	*Tahsil*	*Population*	*Literacy Percentage*
1	Chui Khadan	1,39,447	66.61
2	Khairagrah	1,56,283	76.53
3	Dongargarh	1,74,262	75.66
4	Rajnandgaon	4,53,591	80.35
5	Dongargaon	1,10,730	80.69
6	Chouki	94,365	80.30
7	Mohla	75,968	79.55
8	Manpur	77,165	77.40
	Total	12,81,811	48.77

Source : Census of India 2001.

the most interesting phenomenon like the rest of Chhattisgarh is the high rate of participation of women in agriculture. In agricultural labour, the participation of women is distinctly evident. In fact, female agricultural labourers outnumber their male counterparts. However, in the secondary sector the participation is low.

Land use pattern reveals that 39.66 per cent of the total geographical area comes under net sown area while nearly 1.5 per cent of the area is under fellow. Land not available for cultivation constitutes 21.99 per cent of the total geographical area. The major crops grown in the district area paddy, wheat, maize, and millets. With the development and availability of various agricultural inputs such as irrigation, high-yielding varieties of seeds, fertilizers etc. the cultivated area under these crops is decreasing day by day as farmers are increasingly switching over to more cash crop varieties. The production of different crops varies across tahasils. Nearly 12 per cent of the area under agriculture is irrigated.

There has been an interesting change in the landholding pattern in the district. The trend shows an increasing number of small, marginal and semi-medium landholdings, while the number of medium and large landholdings are declining. The increasing division of joint families into nuclear families and the subsequent redistribution of land among family units may be a major reason for this large fragmentation of lands. This phonomenon, if continues, may lead to expansion of labour class in the district as well as migration to other areas. The total livestock population is 8.98 lakh of which only 30 per cent constitutes productive animals with much of the remaining being unproductive.

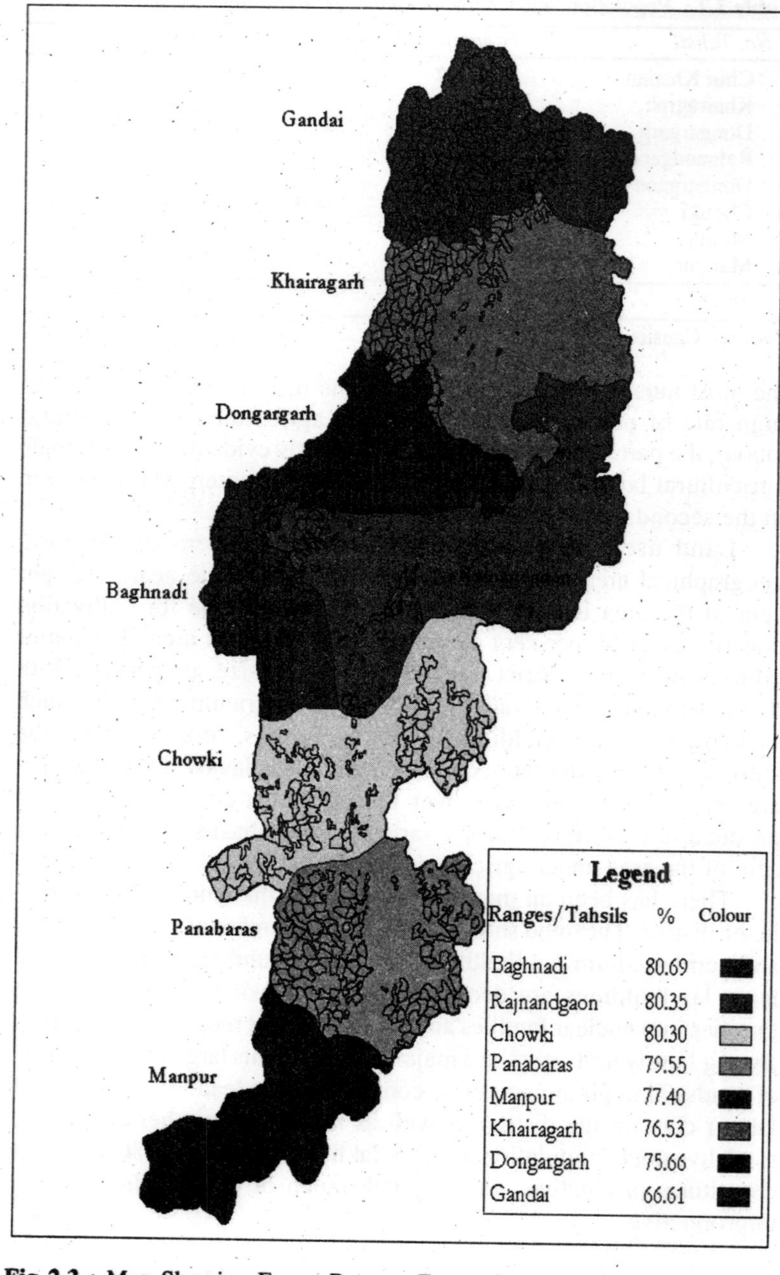

Fig 2.3 : Map Showing Forest Ranges, Forest Compartment and Literacy Percentage

Infrastructure

The district has a fair percentage of good mettalled roads. Each village except the Manpur area has approach roads. With regard to other infrastructural facilities, a lot of improvement has taken place in the period earlier to 1990s. The interior parts of the district woefully lags behind in medical and educational facilities.

Other Factors

Not withstanding population increase and lack of infrastructure, majority of people, mostly living near forests, suffers from poverty. Therefore, all factors indirectly create forces which seemingly operate from the domain of forestry and discourage sustainable development of forests in the district. One of the striking manifestations of such negative forces is rampant uncontrolled fire in the forests. In the fire season, majority of the area burns repeatedly. Grazing is also an inimical factor in the forest environment. Illicit felling and encroachments are other destructive factors. Manpur range is the worst effected range from the point of encroachment. Areas which are near habitations, especially forests adjoining to urban areas are very prone to illicit felling. Bija and Teak are the species susceptible to illicit felling. Apart from this, Naxlite menace, which increased in the plan area since 1991, have severely affected forest management operations.

Disturbance Regime

Natural disturbance is common to forests. Ecologists define disturbance as "any relatively discrete event in time that disrupts ecosystem, community, or population structure and changes resources, subtracts availability or the physical environment (Pickett and White, 1985). To describe a specific disturbance regime and its effects on plant communities, three important parameters must be quantified (Pickett and White, 1985).

1. *Return interval:* The average time between occurrences in a given period.
2. *Severity:* The amount of vegetation destroyed.
3. *Spatial pattern:* Distribution of disturbance effects at various scales, from with-in stand to large landscapes.

Important disturbance agents include felling, NWFP collection, fire, herbivorous outbreak, illicit felling, illicit grazing, and flood. Any increase in human population is a compromise between ecological integrity and society's demand for forest products. Disturbances are a major driver for vegetation change. Disturbance can be defined in terms of changes in physical environment and effects on biota. Disturbances are generally illustrated in terms of a disturbance regime which is the aggregate behaviours of disturbance over long time frames and large areas. Aggregately disturbance behaviour over large scales reduces variability in disturbance. Disturbance regime explains the reasons behind the differences in biological diversity among landscapes.

Forests are highly layered ecosystems. Disturbance can have effect on one layer and leave other layers intact. The magnitude of the disturbance such as fire is usually expressed in terms of intensity or severity, i.e. the amount of live organic matters removed. The timing of disturbances can have profound effects on ecosystem composition and structure. Timing can be viewed in three primary ways: seasonality (time during the year), duration, and frequency which is usually expressed as return interval or rotation time or feeling cycle. The spatial distribution of extent and shape of disturbance are also important features of disturbance regime. Bazzaz (1983) said that what matters in disturbance regime is not what caused the disturbance but what are the nature and consequences of disturbance and how do species and population respond to them over ecological and evolutionary time.

The alteration of forest disturbance regime by humans has four major effects : (1) exclusion of fire from fire dependent ecosystem, (2) reduction in structural and compositional diversity through intensive silvicultural management, (3) conversion of forests into other land use types such as agriculture and (4) alternation of hydrological processes. In most of the cases, human activity has increased fire frequency and intensity causing losses in species diversity and ecosystem productivity. Intensive forest management for timber has altered disturbances regime by increasing the frequency and severity of disturbances. Considered in the above perspective, disturbance regime in the study area presents a variegated picture (Table 2.8). The intensity of grazing is low but the area under disturbance is very large and contrary is the case of main felling. In totality, disturbance factors such as fire, grazing and fuelwood collection leave an adverse impact on development of forests.

Table 2.8 : Disturbance Regime in the Study Area

S.No.	*Name of disturbance*	*Nature of disturbance*	*Area in %*	*Disturbance cycle*	*Final Impact on forests*
1.	Main-felling	Severe but discontinuous	3	Once in 30 years	Mild
2.	Thinning	Mild but discontinuous	10	Once in 10 years	Mild
3.	Grazing	Mild but continuous	100	Once in five days	Severe
4.	Illicit felling	Mild but continuous	5	Once in a month	Mild
5.	Ground Fire	Severe and continuous	70	Thrice in a year	Severe
6.	NWFPs collection	Mild but continuous	100	Once in fortnight	Mild
7.	Fuelwood collection	Mild but continuous	70	Once in week	Severe

Configuration of the Forest Area

The study area comprises different types of topographical features. The district has hilly part in the western side, which runs longitudinally from north to south. The North-western part of the hill is extension of Maikal Ranges. The southern side is the extension of Chhattisgarh Plateau. The area falling in Gandai, Khairagarh and Dongargarh region are mainly hilly in nature. The rest of areas contains hilly, undulating and plain areas. The lowest point in the study area lies at 296 metres above mean sea level.

Description of Forest Area and their Extent

The forests in the study area are spread in an area of 226379 hectare or 2264 sq km. Out of it, reserve forests account for 33 per cent and protected forests account for 67 per cent of the total forests. Best forests of the district belong to the category of reserve forests. However, certain patches of protected forests also come in the category of best-managed forests too. But because of increasing fragmentation of protected forests, especially in those patches, which are situated near village or urban habitation, productivity of the forests is steadily declining.

Reserved forests are situated in Rajnandgaon, Dongargarh, Khairagarh, Baghnadi, Chowki and Panabaras ranges. Protected forests are located in Manpur, Chowki, and Gandai ranges (Fig 2.4). Survey and demarcation of orange area was started in 1996. It was completed in ranges like Manpur, Panabaras, Chowki, Baghnadi/ Dogargarh. In such ranges, out of the surveyed areas, only 743.957 hectare forest area was found suitable for management. This area was added to the existing forests for the purpose of management (Table 2.9).

Table 2.9 : Forest Area in Rajnandgaon district

Sl. No.	*Legal status of forests*	*Number of Compartment*	*Area in hectare*	*Area in %*
1.	Reserved Forests	270	73,719.63	32.6
2.	Protected Forests	679	15,1915.685	67.1
3.	Unclassified Forests (Orange Area)	23	743.957	0.3
	Total	972	226379.272	100

Note : There revenue forest and the forest area leased to Chhattisgarh Forest Developement Corporation were not taken into account.

Source : Working Plan of Rajnandgaon, 1999.

Rights and Concessions

The notifications for Reserve Forests of the study area were issued in the years 1878, 1949 and 1958. After the year 1958, no forest was notified as reserved forest. The forests of Gandai and Manpur were notified as protected forests in the years 1949 and 1958.

The forests are burdened with no rights. However, the Government has given to local people certain privileges and concession. *Wazib-ul-Urg* was an instrument through which Nistar facilities were provided to all local people of the study area. This arrangement continued till the year 1956. In the year 1956 the existing Nistar arrangement was reviewed and the study area was divided into two zones on the basis of their proximity to forests. In the zone situated far off the forests, the privileges were withdrawn, but Nistar depots were opened to extend the Nistar facilities at concessional price. In the zones situated near forests (*i.e.*, with the range of 5 km from the boundary of forests) privileges continued along with Nistar facilities provided at Nistar depot. The figures in Tables 2.10 and 2.11 present the demand and supply of Nistar material in the study area.

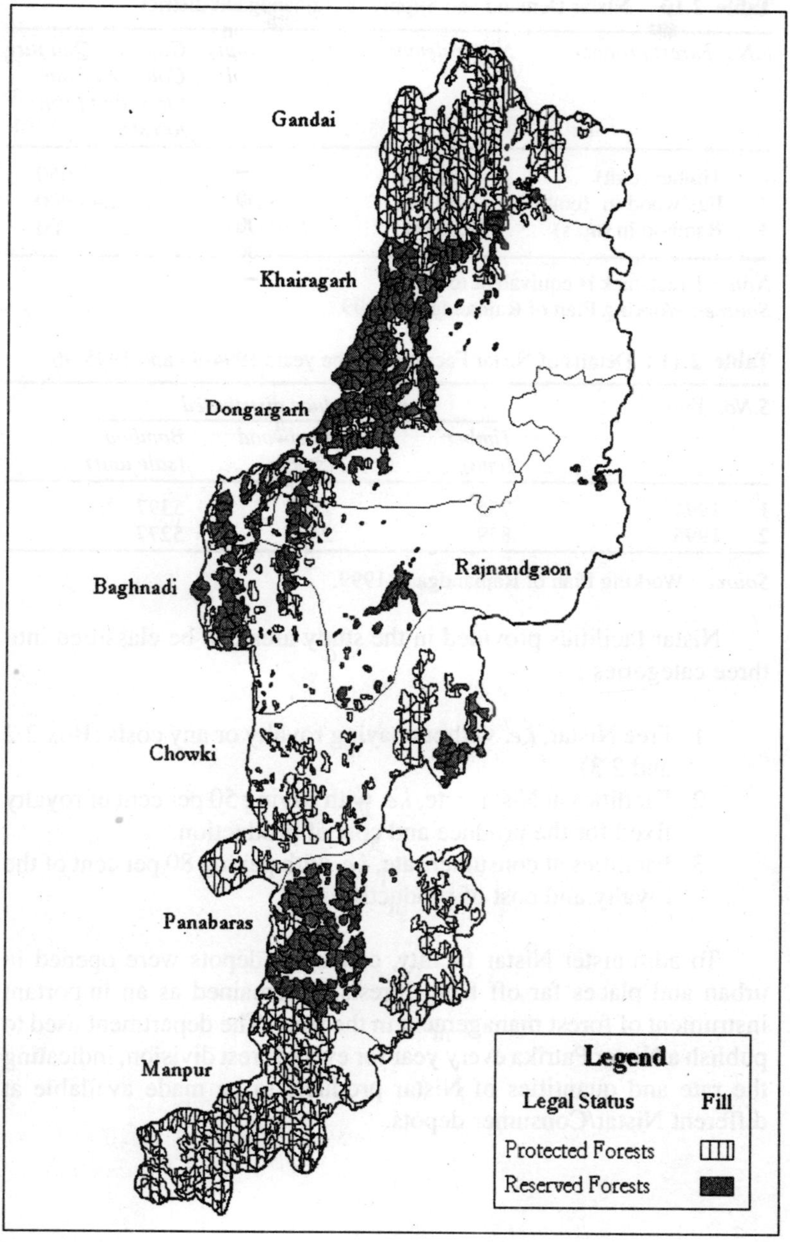

Fig 2.4 : Map Showing Legal Status of Forest Area in the Study Area

Table 2.10 Nistar Demand and Supply in Rajnandgaon district

S.No.	*Forest produce*	*Nistar demand*	*Nistar supply*	*Gap : the Quantity Collected from forest and farm forests*
1.	Timber (cmt)	22,000	950	21,050
2.	Fuelwood in (cmt)	2,50,000	6,000	2,44,000
3.	Bamboo in (no's)	32,00,000	2,50,000	29,50,000

Note : 1 fuelstack is equivalent to 1 cmt.
Source : Working Plan of Rajnandgaon,1999.

Table 2.11 : Details of Nistar Facilities in the years 1994-95 and 1995-96

S.No.	*Year*	*Forest produce distributed*		
		Timber (cmt)	*Fuelwood (stacks)*	*Bamboo (sale unit)*
1.	1994	755	1875	5397
2.	1995	879	2482	5277

Source : Working Plan of Rajnandgaon,1999.

Nistar facilities provided in the study area can be classified into three categories :

1. Free Nistar, *i.e.* without paying royalty or any costs (Box 2.2 and 2.3).
2. Facilities at Nistar rate, *i.e.* with paying 50 per cent of royalty fixed for the produce and cost of production.
3. Facilities at consumer rate, *i.e.* with paying 80 per cent of the royalty and cost of production.

To administer Nistar facility, consumer depots were opened in urban and places far off from forests. It remained as an important instrument of forest management in the state. The department used to publish a Nistar Patrika every year for every forest division, indicating the rate and quantities of Nistar products to be made available at different Nistar/Consumer depots.

Box 2.2 : Nistar and Paidawar : Definition and Meaning

'Nistar' means and includes :

I. Timber of unreserved trees, or reserved trees, expressly sanctioned on that behalf, for making agricultural implements, building new houses, or repairing houses and cattle sheds of the agriculturists;
II. Dry fallen wood not fit for timber;
III. Dry bamboos and green bamboos where specifically mentioned;
IV. Grasses other than Rosa, Khush or Sabai grass;
V. Thorn except those of Khair and Brushwood; leaves including tendu leaves;
VI. Bark (Bakhal) of unreserved trees; and surface boulders, murum, sand and clay.

'Paidawar' means and includes all edible roots, fruits and flowers, naturally excluded gum except the gum from Kulu trees, honey and wax.

'Occupational Nistar means Nistar required for the purpose of carrying on an occupation as a means of livelihood.
Agriculturalist residing in a village shall be permitted to obtain either free of charge or on payment their Nistar and Paidawar requirement from the protected forests to which they have been or may be attached.

Box 2.3 : Free Nistar

Free Nistar includes :

- Dry and fallen twigs in forest
- Small timber
- Poles and small timbers for agricultural implements are provided from Nistar/Consumer depot.

At present, each family is entitled for 16 poles, but it is based on availability. The requirement for agricultural implements is separate. Each agricultural household is entitled to 5 (0.14 cmt) timber every year. Likewise, there is also a provision for providing best part of the trees (Khod) for carpentry at concessional price.

Bamboo

Bamboo remains an important forest produce in the district. People in the rural areas uses it for housing, (making and repairing), basket making and many other purpose. Previously, bamboo used to be collected under Nistar only. But in the year 1989, the Government of Madhya Pradesh issued fresh guidelines for distributing bamboo to villagers. The provision to distribute quantity of bamboo per household in every year is given in Table 2.12. Apart from this, people of the villages away from forests are also entitled to forest produces at consumer price.

Table 2.12 : Provision of Bamboo for Nistar

Sl No.	*Types of Beneficiaries*	*Bamboo Quantity in numbers*
1.	Agricultural Household	250
2.	Basod Household	1500
3.	Pan Bareja	1000
4.	Horticulturists	5

Source : Working Plan of Rajnandgaon, 1999.

Beside this, there is a provision of Nistar to people affected by fire or flood. Every household gets 50 bamboos and 18 poles in case of such calamity. This facility was extended to people from the year 1977. Grazing in the forests is another privilege extended to people living near forests. The Government of Madhya Pradesh issued new grazing rules in the year 1986 to control the unregulated grazing in the forests. In these rules, grazing is allowed on the basis of "carrying capacity" of the forests. This facility is free up to 20 cattle units, but afterwards grazing requires fees. Goat and camel are restricted in the forests. The villagers listed for grazing are given priority while giving grazing permit. In order to make grazing sustainable, the rules require the conservator to constitute grazing units in the division. Each such unit excludes the overgrazed areas. Every year, a survey is required to estimate and demarcate fresh grazing units in the forest division. But, no such survey has been conducted in the study area. It has also been found that conservator is not constituting any grazing units in the division. There is also no effort on the part of administration to control grazing in the forests.

Forest Resources

Forest resources of the district present a rich picture. More than 33 per cent of the geographical area of the district has forest. Apart from this, the forest is rich in productivity as well as diversity. In the ensuing paragraph, we shall discuss extent, legal status, forest types, composition, tree dominance profile, volume distribution, growing stock profile, financial value of forests, etc.

Forest Areas

The district is rich in forest resources. As per the records of the forest department, the district has about 2665 sq. km. of forests. An area of 404 sq. km. of forests is being managed by Chhattisgarh Forest Development Corporation. The rest of the area accounting 2261.295 sq.km. is being managed by forest department (Table 2.13). The area under teak forest constitutes about 8 per cent and mixed forests 82 per cent of total forests. In certain hilly pockets A Gandai, Kheiragarh and Dongargarh salai forests are found as local group.

Table 2.13 : Forest Area on the Basis of Stock Mapping

S. No.	*Composition*	*Forest Area in 1997*	
		Forest Area in Hectare	*In Percentage*
1.	Teak Forest		
	Dense	17031.713	7.53%
	Degraded	366.737	0.16%
	Total	17398.450	7.69%
2.	Mixed Forest		
	Dense	161755.524	71.53%
	Degraded	24311.653	10.75%
	Total	186067.177	82.28%
3	Blank Area & Plantation	18011.578	7.97%
4	Encroachment	4451.540	1.97%
5	Nala, inundated area, other	200.743	0.09%
6	Bamboo	61499.291	
	Total	226129.488	100.00%
7	Forest with Chhattisgarh Forest Development Corporation	40398.072	
	Grand total	266527.56	

Source : Working Plan of Rajnandgaon, 1999.

About eighty per cent of the forest area has dense forests. Ten per cent of the forest area is degraded and open (Fig 2.5). The rest constitutes the blanks, plantation and other such areas. The stages of degradation varies from site to site. From the point of forest site quality 40 per cent of the area belongs to site quality IVA and 25 per cent belong to site quality III. The rest belongs to site quality IVB and VA (Table 2.14 and Fig 2.6). From the standpoint of legality 58 per cent of the forests belongs to protected forests and the rest belongs to reserved forests. Reserved forests are located in ranges such as Khairagarh, Rajnandgaon, Donargarh, Baghnadi, Chowki, and Panabaras. Protected forests are located in all ranges of the district (Table 2.15).

Table 2.14 : Forest Area on the Basis of Stock Mapping in detail

Density	*M.P. Site Quality*	*Forest Area in Hectare*			*Percentage*
		Teak	*Mixed*	*Total*	
Dense Forest Above 0.4 Density (Stocked)	III Quality	8946.474	46834.646	55781.120	24.66
	IV A Quality	7262.142	83146.423	90408.565	39.98
	IV B Quality	782.209	31142.624	31924.833	14.12
	V A Quality	40.843	631.831	672.674	0.30
Total		17031.713	161755.524	178787.237	79.06
Degraded Forest 0.2 to 0.4 Density (Under Stocked)	III Quality	49.176	75.428	124.604	0.06
	IV A Quality	133.341	5816.523	5949.864	2.63
	IV B Quality	148.090	13292.542	13440.632	5.94
	V A Quality	36.130	5127.160	5163.290	2.28
	Total	366.737	24311.653	24678.390	10.91
Grand Total (Stocked + Under Stocked)		17398.450	186067.177	103465.627	89.97
Blank Area		–	–	7629.857	3.37
Plantation		–	–	10381.721	4.6
Other		–	–	4652.283	2.06
Grand Total				226129.488	100.00

Source : Working Plan of Rajnandgaon, 1999.

Table 2.15 : Legal Status of Forest Area

S. No.	*Legal Status of Forest*	*Area under Forest Division (in hect.)*	*Area under FDC* (hect.)	*Total (in Hectare)* (3+4)
1	RF	73719.630	21021.298	94740.928
2	PF	151915.685	19376.774	171292.459
3	Undemarcated	27068.99	–	27068.99
	Grand Total	245369.189	40398.072	293102.377

Source : Working Plan of Rajnandgaon, 1999.

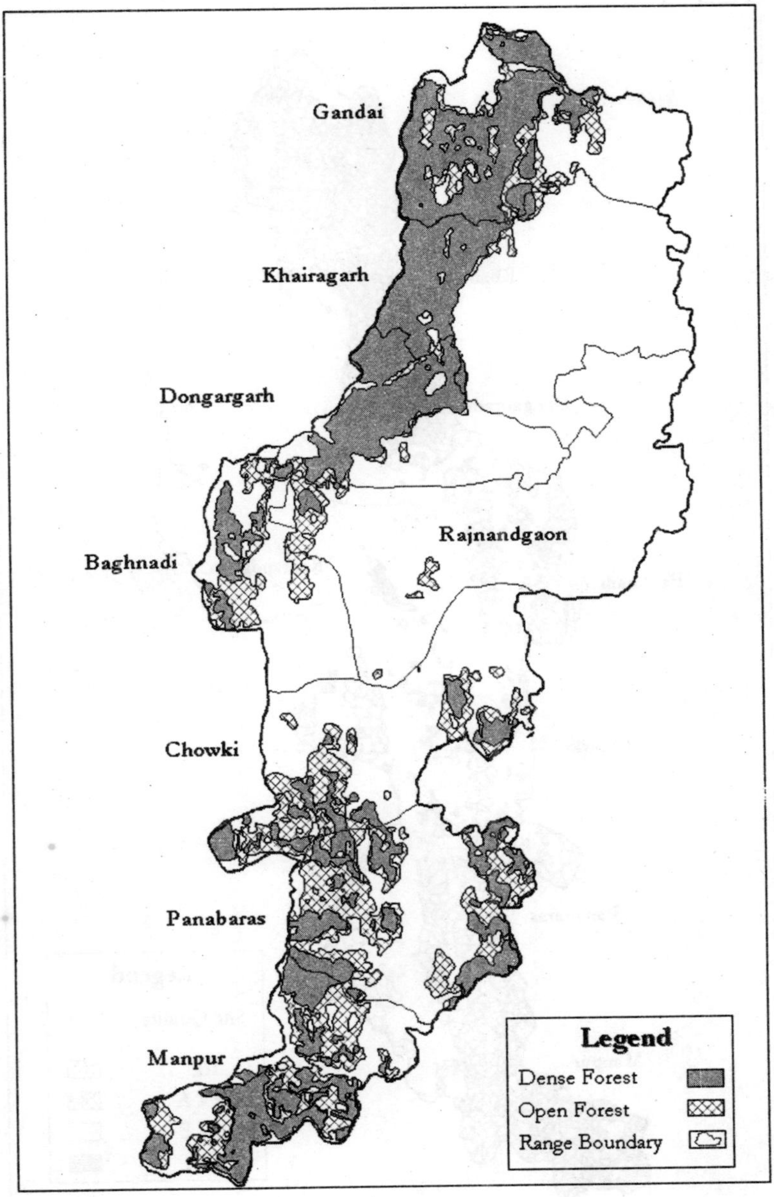

Fig 2.5 : Map Showing Dense and Open Forest Area in the Study Area

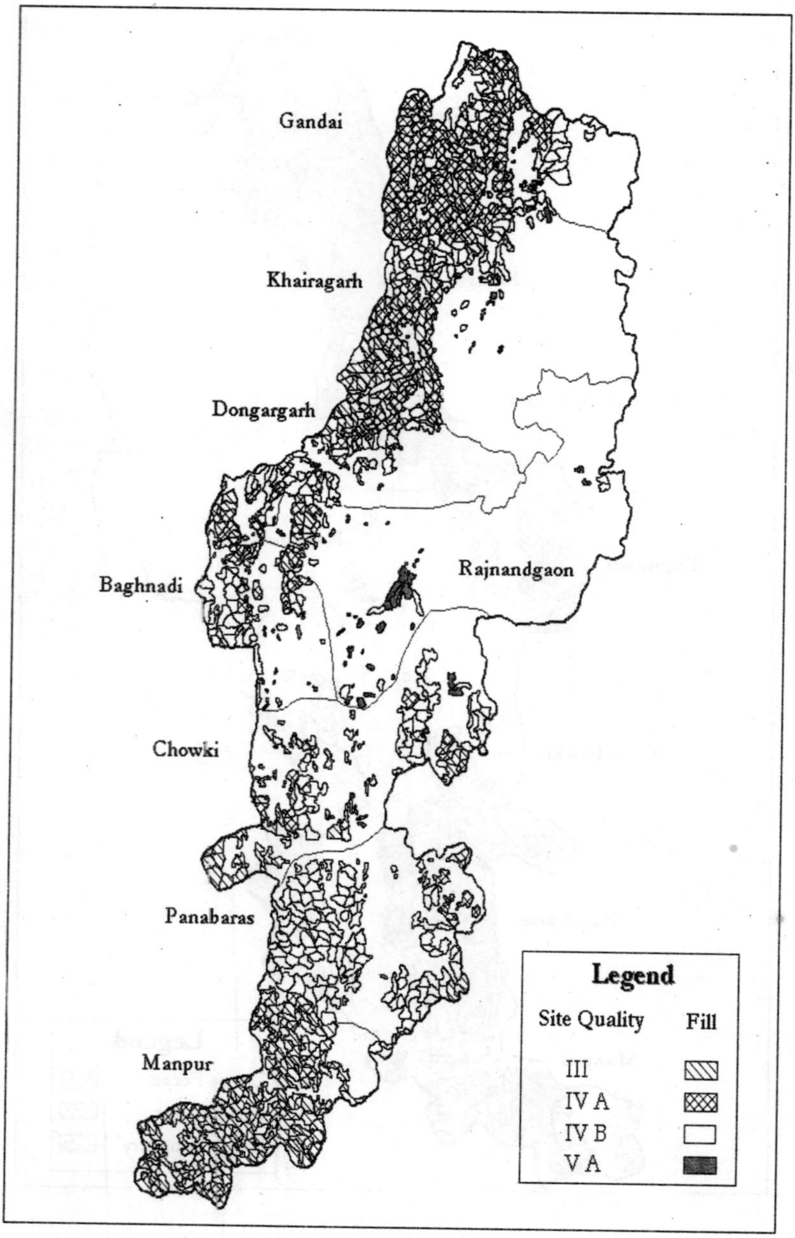

Fig 2.6 : Map Showing Site Quality-wise Forest Area

There are eight ranges in the division. Gandai has larger forest area than any other range (Table 2.16). Panabaras constitutes the smallest range. Manpur is the most productive range having forests of site quality III. Though it has considerably less anthropogenic pressures on forests, but because of productive sites and poverty stricken population, the range is vulnerable to encroachment from tribal population.

Table 2.16 : Range-wise Distribution of Forest Areas

S. No.	*Range Name*	*R.F. (in hect.)*	*P. F. (in hect.)*	*Total Forest Area (hect.)*	*Settled Encroa-chment (hect.)*	*Actual Forest Area (hect.)*
1.	Gandai	-	59632.900	59632.900	341.994	59290.906
2.	Khairagarh	26787.63	5426.810	32214.440	–	32214.440
3.	Rajnandgaon	6650.35	2467.520	9117.870	–	9117.870
4.	Dongargarh	22704.64	5804.980	28509.620	–	28509.620
5.	Baghnadi	14222.19	6619.480	20841.670	–	20841.670
6.	Chowki	3117.68	18044.480	21162.160	42.323	21119.837
7.	Panabaras	237.14	4313.450	4550.590	0.830	4549.760
8.	Manpur	-	50558.350	50558.350	567.138	49991.212
Total		73719.63	152867.970	226587.600	952.285	225635.315
CGFDC : Panabaras division		21419.04	19985.440	41404.480	1006.408	40398.072
Total Forest Area		95138.67	172853.410	267992.080	1958.693	266033.387

Source : Working Plan of Rajnandgaon, 1999.

Chhattisgarh Forest Development Corporation is a state owned public undertaking. It is engaged in enrichment planting in Panabaras range. The corporation plants teak and bamboo in such areas. Plantation has been raised in that range since the year 1975.

Table 2.17 : Forest Blocks and Compartments

S. No.	*Forests*	*No. of Blocks*	*No. of Compartments*	*Total Forest Area (in hect.)*
1.	R.F. Forest	79	270	73719.630
2.	P.F. Forest	313	679	151915.685
3.	Undemarcated Forest	–	23	743.957
	Total	392	972	226379.272

Source : Working Plan of Rajnandgaon, 1999.

Forest Types

Forest types can be classified on the basis of management as well as ecology. If they are categorized on the basis of management, then they are called management type and if categorized on the basis of ecology, they are called as ecological types.

The focus of management changes with the change in the population density of dominant species. If the proportion of teak in the tree population is high, then, it is called teak forests. Likewise if it is less and miscellaneous trees are dominant, then it is called a mixed forest or miscellaneous forests. Categorised on this basis, three types of forests are found. They are: teak forest, miscellaneous forest and bamboo forest (Table 2.18 and Fig. 2.7 and Fig. 2.8). Teak forest is found in Gandai, Khairagarh and Dongargarh ranges. Rest of the area of the forest division comprises miscellaneous forests. In certain hilly pockets of Gandai, Khairagarh and Dongargarh, salai forests are found as local group. The total area of teak forest is 17397 hectares and of miscellaneous forest is 186068 hectares. Rich bamboo forest is found in Khairagarh and Dongargarh range. However, other ranges except Rajnandgaon range also have scattered patches of bamboo forests. Majority of the bamboo forest in the study area belongs to second and first quality category of bamboo. The total bamboo area in the division is 61499.291 hectares.

Table 2.18 : Number of Trees and Volume per hectare

Sl. No.	*Type of forests*	*No. of trees per hectare*	*Mean volume per hectare in cmt.*
1.	Teak	501.875	65.941
2.	Misc.	405.466	65.987

Note : All trees havving 20 cm. Girth and above measured.
Source : Working Plan of Rajnandgaon,1999.

Ecological Types and Composition

A forest type is defined as a unit of vegetation which possesses broad

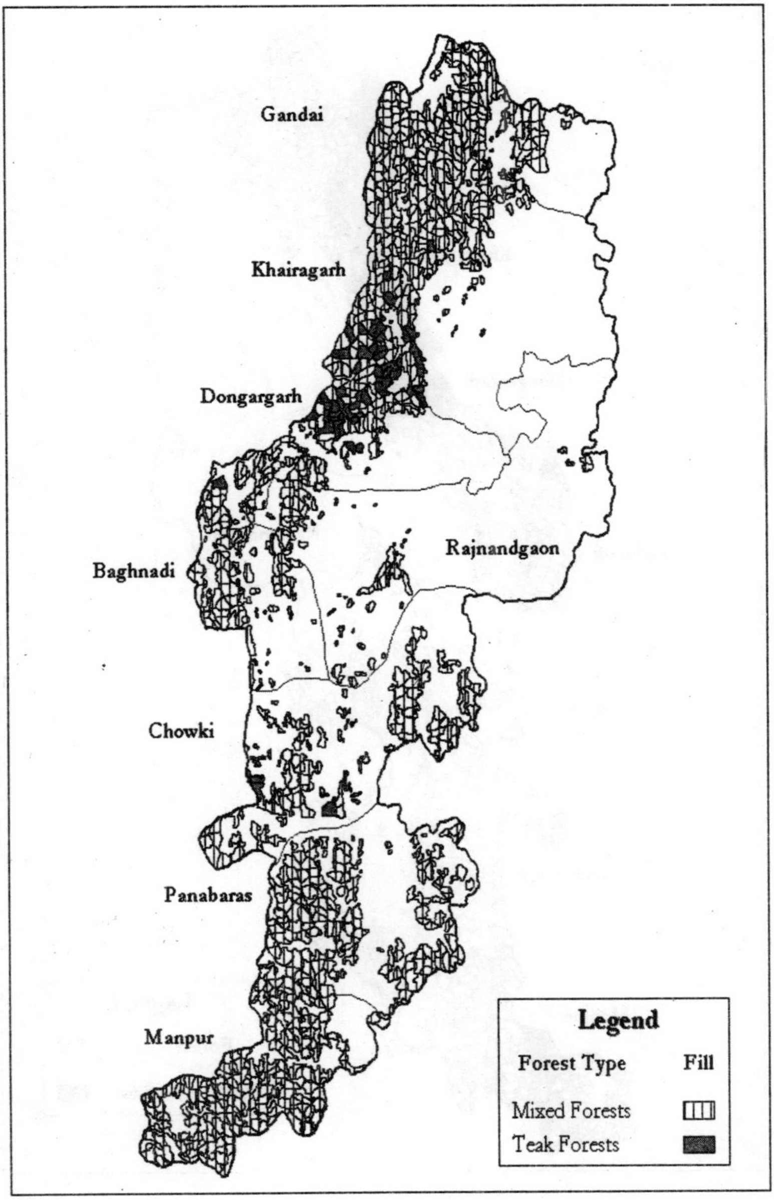

Fig 2.7 : Map Showing Forest Types in the Study Area.

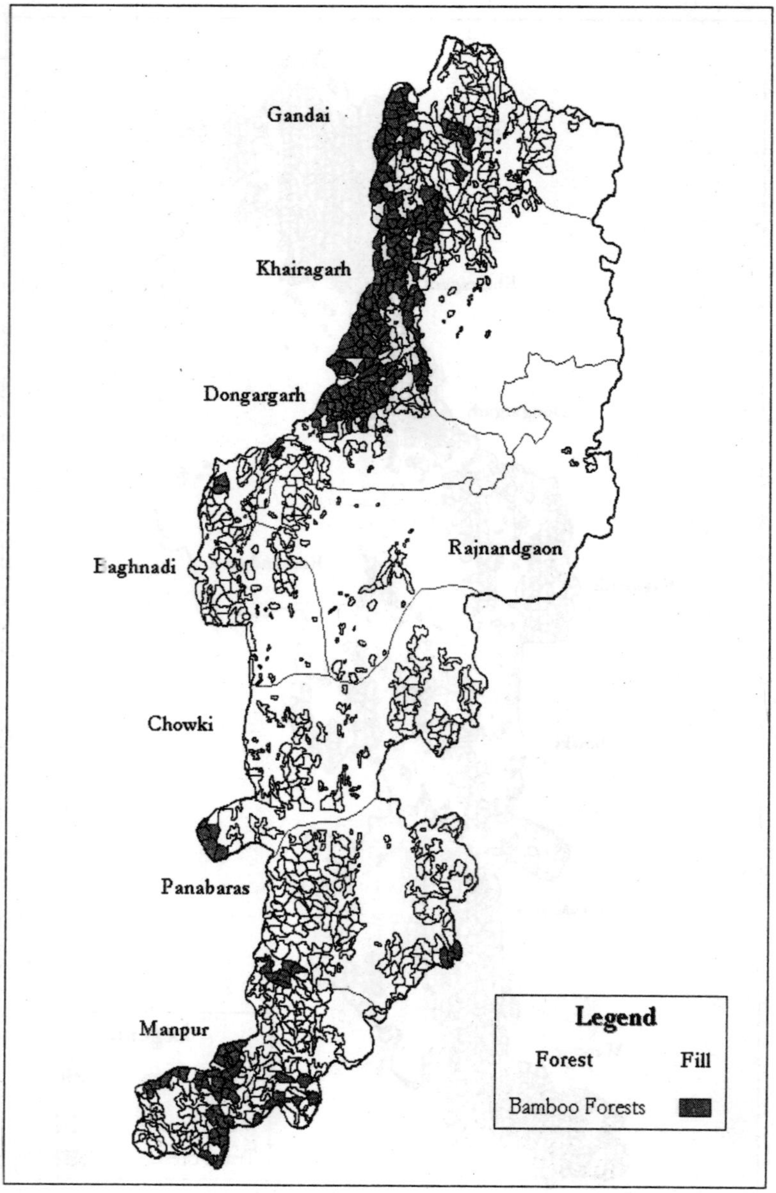

Fig. 2.8 : Map Showing Bamboo Forest in the Study Area

characteristics in physiography and structure sufficiently pronounced to permit of its differentiation from other such units, irrespective of physiography, edaphic and biotic factors. The basic unit of plant ecological study is association which is the largest floristic units consisting of a definite assemblage of species (usually with definite dominants) and a definite habitat. It is simply defined as a community of definite floristic composition within a formation (Bruth-Davy, 1938).

The forests of the study area have been categorized on the basis of Champion and Seth (1964) classification. The forest types are :

1.	Dry Teak bearing Forests	$5A/C_1$
2.	Southern Tropical Dry Deciduous Mixe Forests	$5A/C_3$
3.	Northern Tropical Dry Deciduous Mixed Forests	$5B/C_2$
4.	Boswellia Forests	$5B/E_2$
5.	Dry Bamboo Brakes	$5B/E_9$

Crop in the forest has four distinctive strata. They are : (1) Over wood, (2) Middle Story (3) Ground Story and (4) Grass and climbers.

Dry Teak Forests ($5A/C_1$)

Dry deciduous forests with teak usually form the major proportion of the crop on shallow, porous or stiff clyey soils. Characteristic species of such types are *Tectona grandis, Anogeisus latifolia, Diosypros tomentosa, Hardwickia binata* and other common dry deciduous trees.

This type of forest is found in mountain ranges of Dongargarh, Khairagrh, Baghnadi, Gandai. It is found along with bamboo forests. There are some forests where bamboo is either not found or if found very sparsely. The crown density of such forests varies from 0.4 to 0.7. The main associates of teak are saja, dhawda, bija, haldu etc. According to resource survey conducted in the year 1997 the density of teak trees in such forests is 168 per hect. The total number of trees per hect. is 405.6 trees. In other words, the percentage of teak trees in such forests is 41per cent. It has also been found that the percentage of trees having girth size up to 45cm is at least 65 per cent. The crop is primarily young.

Floristics

- I/II *Tectona grandis, Terminalia tomentosa, Diospyros tomentosa, Anogeisus latifolia, Cheistanthus collinus, Pterocarpus marsupium, Adina cordifolia, Ogeinia oojeinensus*(o), *Delbergia latifolia* (O).
- IIa *Dendrocalmus strictus* (a).
- III *Hellicteres isora, Stobilanthus* sp., *Woodfordia fructicosa.*
- III *Themeda triandra.*
- V. *Bauhinia vahlii.*

Teak is present in high proportion.

Southern and Northern Dry Deciduous Mixed Forests (5 A/C_3 & 5 B/C_2)

This type of forest is found in parts of Khairagarh, Gandai, Baghnadi, Manpur, Chowki, Dongargarh and Rajnandgaon. In areas like Dongargarh, Khairagarh, Gandai, Chowki and Panabaras, such types of forests are found along with bamboo. In other areas, forests are pure miscellaneous either with bamboo or without bamboo. The crown density of such forest ranges from 0.4 to 0.8. According to resource survey conducted in the year 1997, saja trees dominate the crop composition. It has 45 trees per hectare. There are 372 trees per hectare in such forests. The percentage of trees having girth size ranging up to 45 cm is 59. The crop is primarily young.

It differs from the dry teak mainly floristically. Bamboo are often absent or if present they are of poor quality. The most characteristic tree is *Anogeissus latifolia* while *Terminilia* is a typical associate. *Dyspyros tomentosa* is also common. *Choroxylon switenia, Hardwickia binata, Boswellia serrata* and *Soymida febifuja* are very widespread and they are the useful indicators of the ecosystem. *Terminalia tomentoasa* and *Cleistasnthus collinus* predominate in closed canopy on the better soil.

Floristics

- I/II *Terminilaia tomentosa, Cleistanthus collinus, Lagerstroemia coliinus, Lagerstroemia parviflora, Anogeisus latifolia, Choroxylon swietenia, Emblica officianalis, Pterocarpus marsupium, Ougeinia oojeinensis, Lanea cormondelica.*

- IIa *Dendrocalmus strictus* (a).
- III *Gardenia resinifera, Ixora arborea, Hellicteres isora, Woodfordia fructicosa, Indigo ferapulchellla.*
- III *Heteropogan conturtus, Themeda triandra.*
- V Climbers few. *Bauhinia vahlii, smilex.*
- About 15 per cent of miscellaneous forests come under the category of degraded forests, where the crown density has declined from 0.4.

Boswellia Forests (5B/E_2)

Boswellia forests are found on a variety of soils, particularly the dry plateux, steep slopes and ridges and spurs of mountains found in ranges like Dongargarh, Rajnandgaon and Gandai. It is basically an edaphically climax types in dry deciduous forests. It is found in the lataritic rocks of the western Rajnandgaon. The soil has boulders, and it is shallow and dry. Hill tops, ridges and spurs and well drained plateux are the usual sites,where boswellia forests or stands of boswellia trees are found. Boswellia is usually left standing when other trees are felled and it is very fire resistant. It generally regenerates through root suckers. In open forest bowswellia serrata forms an overwood. Sometime pure aggregation of well grown mature trees occur, especially on the tops and upper part of the slopes of the extended Satpura range. It remains leafless from February to May. The percentage of Salai trees in such forests is up to 7 to 10 per cent of the population. The Salai trees are of middle aged.

Floristics

- I/II *Boswellia serata, Cochlospermum retigiosum, Sterculia urens, Lagerstroemia parviflora, Emblica officianalis.*
- III Practically no shrub.
- Vb. Grass very light.

Bamboo Forests

The study area constitutes one of the few rich patches of bamboo forests in the state. The quality of bamboo forests stands in equal footing with best of Asian standard. The bamboo species is Dendrocalamus strictus. Site quality I and II constitute the majority of bamboo areas. They are found in the hilly tracts of Dongargarh and Khairagarh. Apart from this, bamboo is also found in Manpur, Chowki, and Panabaras areas. The extent of bamboo forests in the study area is 61499.291 hectares. This constitutes about 27 per cent of the total forest area. Aggregation of good bamboo forests is generally followed with severe exploitation. This has happened in the study area too. As a result of this, degradation has set in. Presently, the degraded bamboo forests constitute about 15 per cent of the total bamboo forests (Table 2.19).

Table 2.19 : Bamboo Forests under different Site Qualities/Density (in hectare)

S. No.	*Site Quality*	*Dense*	*Medium*	*Degraded*	*Total Forest Area*
1	2	3	4	5	6
1	I Quality	18571.494	2670.333	27.680	21459.507
2	II Quality	14973.894	12938.891	3663.690	30935.935
3	III Quality	431.870	2914.259	5117.180	9103.849
	Total	33977.258	18523.483	8998.550	61499.291

Source : Working Plan of Rajnandgaon, 1999.

From the perspective of site quality, 35 per cent, 50 per cent and 15 per cent of bamboo forests fall in site quality I, site quality II and site quality III area respectively. The average clump size is about 18 culms per clump.

If we analyze the growing stock of bamboo from the point of age computation, then it is found that there is an element of improportionality or imbalance in age distribution. What should have been the balanced crop composition is a moot point here. From the point of crop sustainability the distribution of bamboo culms in all the four age brackets should have been about 25 per cent each. Furthermore, the annual recruitment to first year bamboo stock should have been 25 per cent of the growing stock. But, the ground situation is different. The first year bamboo is only 16 per cent of the total culms (Table 2.22). This is about nine per cent shorter than the annual recruitment target. From the point of sustainability, this is an alarming situation. An imbalance in age distribution is basically the result of uncontrolled fire incidences in the bamboo forests as well as maltreatment of the resources.

There are about 289 clumps per hectare in site quality I area, 201 clumps per hectare in site quality II area and 37.5 clumps per hectares in site quality III areas (Table 2.20).

Table 2.20 : Distribution of Bamboo Clumps and Culms per hectare

S.N.	*Bamboo Forest class*	*Clumps per Hectare*	*Culms per Hectare*
1.	Site Quality I	288.795	3406.594
2.	Site Quality II	200.795	3457.402
3.	Site Quality III	37.5	559.37

Source : Working Plan of Rajnandgaon, 1999.

If we analyze the growing stock from the point of overall crop condition, it is found that 84 per cent of the culms are healthy and the rest are degraded or dry. Therefore, the bamboo crop needs improvement on a sustainable basis (Table 2.21 and Table 2.22).

Table 2.21 : Age Composition of Bamboo Culms

S.No.	*Item*	*No of culms in lakh*	*Culms in %*
1.	1st year Bamboo(Karla)	142	16
2.	2nd year Bamboo(Mahila)	257	28
3.	3rd & 4th year Bamboo(Pakiya)	503	56
	Total	902	100

Note : The total number of clumps in the study area is 51.494 lakhs.
Source : Working Plan of Rajnandgaon, 1999.

Table 2.22 : Condition of Bamboo Growing Stock

S.No.	*Condition*	*No of culms in lakhs*	*Culms in %*
1.	Degraded	74	8
2.	Dry	66	7
3.	Healthy	762	84
	Total	902	100

Source : Working Plan of Rajnandgaon, 1999.

Stem Distribution in Teak and Mixed Forests

Stem distribution plays an important role in nutrient recycling and growing space management in the forests. In normal forests, the distribution of trees of different ages is always normal. That means, if we draw a curve taking girth class in the X-axis and number of trees in the Y-axis, then inverted J-shaped curve will appear. This means, the stem distribution is normal. But, normal forests is only ideal. They do not exist as such. What is feasible is somewhat nearer to this stage.

In teak forests of the study area, the teak dominates the tree population. Next to it, comes saja and garari. Other species are mode, tendu, bija, etc. (Tables 2.23 and 2.24). In the teak and miscellaneous forests, the distribution of trees is not inverted J-shaped. There is a lot of distortion in the shape of the stem distribution.

Table 2.23 : Dominance profile of important tree in teak and mixed forests (No. of trees)

S. No.	*Species*	*Teak Forests*	*Mixed Forests*
1.	Teak	83.542	11.20
2.	Saja	30.000	34.035
3.	Mode	14.999	15.265
4.	Garari	24.658	24.588
5.	Bija	7.083	7.715
6.	Salai	9.241	9.282
7.	Dhawda	11.459	20.303
8.	Others	59.013	131.510
	Total	239.995	256.373

Note : All trees having 30 cm Girth and above measured.
Source: 1997 Resource Survey. (Working Plan of Rajnandgaon:1999.)

Table 2.24 : Stem distribution in teak forests

S. No.	*Girth Class (in cm.)*	*According to 1997 – Resource Survey*	
		No. of Tree per Hectare	*Volume per Hectare(cmt)*
1	2	3	4
1	30-45	96.870	5.806
2	46-60	42.294	4.634
3	61-90	66.458	20.048
4	91-120	22.083	15.085
5	121-150	7.708	9.611
6	Above 150	4.584	7.602
	Total	239.995	62.796

Note : All trees having 30 cm girth and above measured.
Source : 1997 – Resource Survey (Working Plan of Rajnandgaon:1999).

The age structures of the tree population of the forests are dominantly young to middle aged (Fig 2.9). There are also stray patches of matured crop. Majority of the young crop is coppice in origin. But, the middle aged and mature crop is of seed origin.

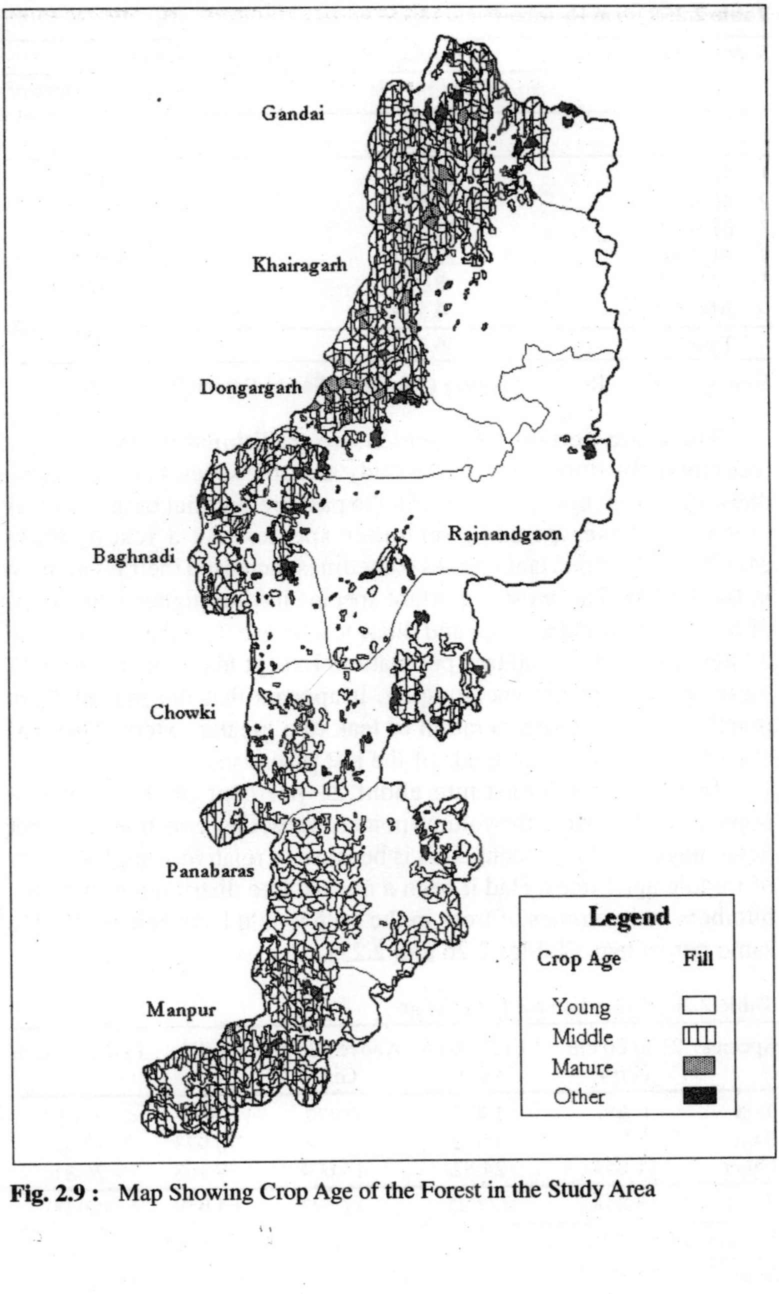

Fig. 2.9 : Map Showing Crop Age of the Forest in the Study Area

Table 2.25 : Stem Distribution in Mixed Forests

S. No.	*Girth Class (in cm.)*	*According to 1997 – Resource Survey*	
		No. of Tree per Hectare	*Volume per Hectare(cmt)*
1	2	3	5
1.	30-45	103.11	6.446
2.	46-60	55.672	6.267
3.	61-90	61.489	18.084
4.	91-120	22.607	15.369
5.	121-150	8.952	10.978
6.	Above 150	4.538	7.232
	Total	256.373	64.375

Source : 1997 – Resource Survey (Working Plan of Rajnandgaon, 1999).

There are certain trees which are traditionally classified as 'economically important'. In the study area, the species are teak, saja, dhawada, tinsa, bija, haldu, etc. In the past preferential treatment was given to these species over other species. As a result, these economically important species sometimes increased their population in the forests. The most important species having higher percentage of population are teak, saja and dhawada. These three species exceed 27 per cent of the total tree population. Except teak, regeneration of these species are not encouraging. It appears that the management practices favoured regeneration of teak to a greater extent. This was one of the management goals of the last plan also.

Teak and saja constitute about 21 per cent of the total tree population. But, from the volume point of view, they constitute a higher percentage, i.e. 23 per cent. This is because of relatively high number of middle aged trees. Had it been a normal tree distribution, both the numbers and volumes of trees in the crop would have constituted the same percentage (Tables 2.26 and 2.27).

Table 2.26 : Mean Volume Per Hectare

Species	21 to 60 cm Girth	61 to 120 cm Girth	Above 120 cm Girth	Total	Percentage %
Teak	1.206	1.833	0.879	3.918	6.12
Saja	1.292	5.972	3.709	10.973	17.07
Other	11.044	24.82	13.004	49.168	76.81
Total	13.842	32.625	17.592	64.059	100.00
Percentage	21.61	50.93	27.46	100.00	

Source : 1997-Resource Survey (Working Plan of Rajnandgaon, 1999).

Table 2.27 : Stem distribution of economically important trees

Sl. No.	*Species*	*No. of trees in different Girth Class per Hectare*										
		11 – 20	*21 – 30*	*31 – 45*	*46 – 60*	*61 – 90*	*91 – 120*	*121 – 150*	*151 – 180*	*Above 180*	*Total*	*Percentage*
1.	2	3	4	5	6	7	8	9	10	11	12	13
1	Teak	8.473	13.083	8.728	3.134	3.608	1.020	0.419	0.128	0.091	38.684	9.64
2	Saja	1.822	10.441	9.308	6.469	10.277	4.136	1.713	0.656	0.310	44.862	11.18
3	Dhawada	2.150	8.345	6.487	4.391	5.594	1.658	0.565	0.091	0.073	29.354	7.31
4	Tinsa	0.310	1.294	2.351	1.640	1.968	0.346	0.128	0.036	0.000	8.073	2.01
5	Bija	0.091	0.893	1.130	0.620	2.679	1.786	0.765	0.219	0.200	8.383	2.09
6	Haldu	0.128	0.182	0.310	0.146	0.200	0.237	0.091	0.073	0.018	1.385	0.34
	Total	12.974	34.238	28.044	16.400	24.326	9.183	3.681	1.203	0.692	130.741	32.571

Source : Working Plan of Rajnandgaon:1999

The study area indicates a large range of variability in terms of site quality too. The site quality ranges from III to Va quality. The most productive site quality in the study area is III quality, which supports 84 cubic meter of wood per hectare where as the least quality *i.e.* Va supports only 24.4 cubic meter of wood per hectare (Table 2.28).

Table 2.28 : No. of Trees and Volume Per Hectare on the basis of Site Quality

Site Quality	*No. of Trees per Hectare*	*Volume per Hectare (in cmt)*
II Quality	–	–
III Quality	363.310	83.800
IV a Quality	462.667	70.008
IV b Quality	419.171	54.343
V a Quality	430.000	24.407

Source : Working Plan of Rajnandgaon, 1999.

From the age point of view, the crop is not harmonized. There is dominance of young trees, which constitutes 80 per cent of the teak forests and 76 per cent of mixed forests. The middle-aged trees constitute about 18 per cent in teak forests and 21 per cent in mixed forests. Likewise, the matured tree is only 2.5 per cent and 3.5 per cent of teak and mixed forests respectively (Table 2.29). Possibly, this situation has arisen because of increased removal of mature trees over a period of about three felling cycles, which cover about 75 years. This might have not only resulted into sharp decrease in matured trees but also it promoted young trees.

Table 2.29 : Volume of Trees Per Hectare in different Crop Age

S. No.	*Forest Type*	*No and Volume of Trees Per Hectare in different age classes*						*No and Volume of Trees Per hect. (in cmt)*	
		Young		*Middle*		*Matured*			
		No	*Volume*	*No*	*Volume*	*No*	*Volume*	*No*	*Volume*
1	Teak	401.047	13.594	88.541	35.134	12.292	17.213	501.875	65.941
		79.9%	20.6%	17.6%	53.2 8%	2.5%	26.11%	100%	100%
2	Mixed	307.875	14.324	84.096	33.453	13.495	18.210	405.466	65.987
		75.9%	21.70%	20.8%	50.7%	3.3%	27.6%	100%	100%

Note : All trees having 21 cm. girth or above recorded.
Source : Resource Survey (Working Plan of Rajnandgaon, 1999).

The crop volume per hectare in dense forests is very high in comparison with degraded and blank areas. The reason is very obvious.

Because of better site quality and presence of good coppicers, healthy young and matured trees are found. The middle aged trees are generally illicitly felled. The number and volume of trees in degraded forests speak volume about the scale of degradation (Table 2.30).

Table 2.30 : Crop Profile in Dense and Degraded Forests

S. No.	*Density*	*No. of Trees per Hectare*	*Volume per hectare (in cmt)*
1.	Above 0.4 (Dense Forests)	467.323	71.052
2.	0.2 to 0.4 (Degraded Forests)	173.537	43.247
3.	0 to 0.2 (Blank Area)	19.531	8.263

Source : Resource Survey (Working Plan of Rajnandgaon, 1999).

Among the ranges, Manpur, Khairagarh and Chowki are the most productive forest ranges in the study area. Manpur and Chowki ranges lie in the south of the study area. The Khairagarh range is located in the north of the study area. In all these ranges the soil is deep and sandy loamy in texture. The drainage is very good. Manpur range, receives relatively high rainfall. The volume per hectare is 77.5 cmt. This speaks volume about the productivity potential of the range (Table 2.31). The ranges such as Baghnadi, Rajnandgaon and Gandai are the least production ranges.

Table 2.31 : Number of Trees and Volume per hectare range-wise

S. No.	*Density*	*No. of Trees per Hectare*	*Volume per hectare (in cu. mt.)*
1.	Dongargarh	461.280	53.825
2.	Panabaras	454.688	51.563
3.	Gandai	437.787	54.089
4.	Khairagarh	424.185	74.026
5.	Baghnadi	410.197	49.677
6.	Manpur	384.889	77.563
7.	Rajnandgaon	300.962	37.369
8.	Chowki	253.427	60.354
	Total	402.077	64.059

Source : Resource Survey (Working Plan of Rajnandgaon, 1999).

Growing Stock

Growing stock in economic parlance can be called as natural capital of the forests. Sustainability of natural resources depends on how we use the growing stock. We should manage and use our natural capital in such a way that while reaping the environmental, social and economic benefits, we should not diminish our forests in terms of extent, quality and diversity.

The growing stock of the forests is estimated at 13.1 million cubic metres. Out of it teak forest constitutes 1.06 million cubic metre and mixed forests constitutes 12 million cubic metres. It means, teak forests constitute about 13 per cent and mixed forests 87 per cent of the total growing stock. Economically important species like teak, bija, saja, tinsa, dhawda and haldu constitute 33 per cent of the growing stock. Again, saja alone constitutes 11 per cent of the total tree population of the forests.

Financial Value of Growing Stock

Out of the growing stock 13.83 per cent constitute trees which produces non-wood forest products having a reasonable market demand. Under this category, species like Tendu, Amla, Achar, Harra, Mohua and Baheda comes. If we combine both timber and non-timber trees having some economic importance, then 47 per cent of the trees have economic importance and the rest has only ecological importance.

Given a financial value, the growing stock accounts for Rs. 63,222 millions. Out of it, timber and bamboo accounts for Rs. 61,959 millions and Rs.1263 million respectively. Again analysing the financial values from the perspective of working circle, the growing stock in SCI working circle constitutes the highest financial value (Table 2.32).

Table 2.32 : Financial Value of Tree and Bamboo Growing Stock

S.No.	*Working Circle*	*Value of Growing stock in Rs. Millions*			
		Teak Crop	*Mixed*	*Bamboo Crop*	*Total Value*
1.	SCI	6530.8	28221.6	1003.4	35755.8
2.	Improvement	3688.7	19321.9	252.8	23263.4
3.	RDF	663.8	3532.1	6.5	4202.4
	Total	10883.3	51075.6	1262.7	63221.6

Note : Selection-cum-improvement (SCI), Improvement and rehabilitation of degraded forests are the three main working circles into which the forest compartments were allotted.

Source : Resource Survey (Working Plan of Rajnandgaon, 1999).

3

Past Forest Management: An Appraisal

A management plan or working plan is meant for a definite period such as 10 years or 15 years. After the end of that period, a revised plan will be at place. Therefore, revision of the plan commences before the plan ends. During the process of revision, evaluation of the plan is undertaken. Evaluation is an important management function. The impact of the last plan on different aspect of forests is critically evaluated and on the basis of this, new plan prescriptions are evolved.

In this chapter a review of the last forest working plan pertaining to the study area, *i.e.* the district has been done. The salient features of results of past management have been brought forth. Beside this, a working circle-wise evaluation has also been done to find out the impact of management practices on forest crop. In order to look at the impact of last plan's prescription on a holistic basis, an impact analysis has been done with respect to different facets of forest ecosystem.

Management Planning over the Years

Forest management in the study area has already completed more than 100 years. It constitutes a chequered history. Initially, *i.e.* in the last leg of eighteen century or in the first decade of the nineteenth century, there was more regulation and less management. Then came deregulation and system of intensive felling. This phase continued till the close of the Second World War in 1939. After this, a period of consolidation, conservation, regulation and scientific management began which continued till date. Therefore, looking at the varying patterns of forest management in different phases, one can easily recognize the chequered character forest management history in the study area.

To understand the evolution of forest practices, the forest history of the district can be divided into three phases: (1) pre-war and war period; (2) pre-independence period, and (3) post-independence period. The pre-war and rear period is characterized by license felling. There was very less anthropogenic pressure on the forest. Therefore, in spite of unscientific management, the forest continued its vitality and resourcefulness. Thereafter, came the war period that spanned from 1919 to 1939. This period is characterized by very heavy systematic felling of middle aged and matured trees of valuable species. As a result, the stock of matured and to some extent middle-aged crop declined sharply. They were basically carried out to meet the war requirements. Very heavy selective fellings of the crop remained a conspicuous feature of this period. Post war period, began in 1939, interlaced with different phases of restoration and exploitation. With the attainment of independence and enunciation of a new forest policy in 1952, the objective of forest management changed slightly. New plantation of teak was taken up on a wide scale by clear felling rich forests. The pace of selective felling of valuable matured crop declined but till that time the stock was already far below the desired mark. As a result, the crop bears a predominantly middle-aged character.

The forests of the district can be spatially divided into three entities. First, it is the reserve forests spread over three ranges such as Khairagarh, Rajnandgaon and Dongargarh. The objective of management in this region was to produce large sized timber. These areas are basically teak bearing forests. These areas were worked under selection system and selection-cum-improvement system. Very heavy felling took place in such forests in the past. Second, it is the protected forests found in all ranges, but more dominantly in range like Gandai, Manpur and Chowki. These regions were under the control of local Zamindars in the past. During this period forests were put under either very loose regulation or systematic management scheme. This state of affairs continued till the independence and after it, such forests were acquired by the provincial government of CP and Berar. After their acquisitions, the forests were subjected to systematic scientific management. Except in the range of Manpur and some part of Gandai, these forests are basically located near villages and are of low site quality. The third is the orange area or undemarcated forests scattered in the outskirts of protected forests. Major chunk of forests basically exists on records. After acquisition of large chunks of protected forests, and more importantly, of forests away from villages, the strip of forests adjoining the village boundary was left out to remain with the revenue department for the whole of period of second part of the last century. As a result, these forests were left unmanaged for about five decades. Over the year, a large chunk of such forests was subjected to illegal encroachment for purpose of cultivation. Now, such forests are again added to forest department's record for management.

The prevailing socio-economic condition, demand on forest produce, especially logs for ship building and political system present in that part of time shaped the contours of past forest management in the study area. As already said, the history of forest management can be divided into three distinct phases. These phases are:

1. Pre-war and war period	Up to 1919 & 1919 to 1939
2. Post-war period	1939-1950
3. Post-independence period	1950-2000

Pre-war and War Period

Before the year 1893, felling of trees in the forests was licence driven. The only requirement for felling of tree was the licence. Licences were used to be given to people without consideration of the size or condition of the tree. Therefore, during the period before 1893, licence used to regulate the felling in the forests. But, after 1893 there occurred a change. Systematic fellings of trees began in forest areas such as Kheiragarh, Dongargarh, Rajnandgaon, Bhagnadi, Chowki, Panabaras, and Manpur. It was primarily based on observation of characteristics of trees on ground. It was initiated by a European Officer Mr. A.W. Blunt. This symbolized the beginning of scientific management of forests in the study area, (Table 3.1). Most of the forests of the district were under the proprietary control of erstwhile Zamindars of Gandai, Khairagarh, Rajnandgaon, Chowki, Khuji, Dongargarh. These Zamindars gave a sizable chunk of good forests to Bengal Trading Company on lease in the year 1905. The lease continued till 1932. After expiry of lease, some of the Zamindars appointed forest officer Mr. Ghanhor to survey the forests and to prepare a management plan. The said forest officer prepared the plan and it remained effective till 1953. The said plan promoted extraction of big sized timber from the forest during the war period. Therefore the plan was more of exploitative and regulatory in nature then developmental. The said plan, to a great extent depleted the matured crop of species such as Teak, Saja and Bija from the forest in one hand. In the other, it failed to regulate the anthropogenic pressure on forest.

Post-war Period

Since there was a heavy depletion of large sized timber from the forests, especially Teak, Saza and Bija, during the post-war period, attempt was taken to rehabilitate the forests with large sized timber. Towards this end, silvicultural system like selection-cum-improvement felling was adopted. This process of restoration still continues today, *i.e.* after about sixty years.

Post-Independence Period

The post independence period saw a new forest policy in 1952. With this the focus of forest management got widened. There was an emphasis on bringing all forests under the fold of working plan. Successful attempt was also made to bring more and more forests, which were with the Zamindars as village forests, under the proprietary control of provincial government.

Then the Gandai part of the study area along with Kawardha area was taken up for preparing working plan. The plan for Gandai was formulated by Mr. A.C. Nigam. It lasted up to 1968. The said plan was basically a working scheme. After the expiry of that plan two working schemes took its place. First was formulated by Mr. Gulati and second was formulated by R.K. Sukla. After it, the Gandai range along with Kawardha was covered by Mr. P. Sahu's plan. This became operational from the period 1982 to 1999. The plan was only implemental in Gandai range of the study area.

In the post-independence period, apart from the efforts made towards consolidation of forest areas, the forest management strategy, based on the new forest policy was revised. Management tried to focus on the following objectives :

- Improvement of the growing stock.
- Increase of middle aged and matured crop in the growing stock.

- Enrichment planting with teak plantation in areas having site quality III.
- Expansion of natural teak area.
- Exploitation of Salai gum and wood.
- Management of bamboo forests.

Table 3.1 : Time Line of Forest Management Planning in Pre-Independence Period

S.No.	*Management Planning*	*Period*
1.	Regulated Felling through Licence	Upto 1893
2.	Systematic Felling	1893 to 1904
3.	Regulation of Systematic controlled felling by Bengal Trading Co.	1905 to 1922
4.	Working Scheme	1923 to 1956
5.	Working Plan/Working Scheme	1957/1951

In order to achieve the aforesaid management objectives, the following working circles were usually constituted in the working plan/schemes :

- Selection-cum-Improvement.
- Coppice with Reserve.
- Bamboo (Overlapping) Working Circle.
- Miscellaneous Working Circles.

In the post-independence period, a renowned forest officer of CP and Berar, Mr. Waheed Khan prepared a new working plan for the reserved forests. It remained effective from 1952 to 1960. The said plan prescribed five working circles for management of forest (Table 3.2).

In the year 1960, A new working plan was prepared by Mr. Argal. It remained effective till 1977-78. Mr. Argal prescribed five working circles as made by Mr. Waheed Khan (Table 3.3). Mr. Argal discontinued the Lakh and Kullu (overlapping) working circle because of dwindling resources of Kullu and Kusum trees in the forests. Regeneration of Kullu appeared to be a major problem. Apart from this, selection-cum-improvement working circle and improvement working circle were merged together to form one SCI working circle. This was done because the study area encompassed dominantly middle-aged crop, which required silvicultural treatment of improvement.

Table 3.2 : Profile of M.A Waheed Khan's Working Plan : 1952-60

Sl. No.	*Working Circles*	*Proposed silviculture system*	*Forest to which applied*
1.	Selection-cum-Improvement	Selection and improvement felling	Dominantly matured age crop: Teak, Sal and mixed forests
2.	Improvement	Improvement felling	Dominantly young and middle aged crop
3.	Coppice with Reserve	Coppice with retention of seed reserve trees	Young aged coppice crop near habitation
4.	Teak Plantation	Clear felling with teak plantation	Site quality III areas suitable for teak plantation
5.	Bamboo Working Circle	Bamboo cutting with four year cutting cycle	Bamboo forests
6.	Lakh and Kullu (Overlapping) Working Circle	Lakh cultivation and Kullu gum tapping	Mixed and Salai forest

Table 3.3 : Profile of Argal' Working Plan : 1960-78

S.No.	*Working Circle*	*Silvicultural system*	*Forest to which applied*
1.	Selection-cum-improvement	Selection and improvement felling	Dominantly matured age crop
2.	Coppice with Reserve	Coppice	Young aged coppice crop near habitation
3.	Plantation	Clear Felling with planting	Site quality III areas suitable for plantation
4.	Bamboo Overlapping	Bamboo felling with four year cutting cycle	Bamboo forests
5.	Miscellaneous	Moratorium on felling or regulated selection felling	Forest compartments containing forest villages

After the expiry of Argal's working plan on 1978, a new working plan by Mr. I.C. Goel came into operation. This plan became effective from the year 1978 to 1999. This was operational in whole of the study area excluding Gandai range. In both of the plans, the working circles, which were in vogue, are presented in Table 3.4. Then Kailash Chandra Bebarta, the writer himself, revised the existing working plan to manage the forests of the study area, *i.e.* the whole of Rajnandgaon district.

Table 3.4 : Profile of Goyal's Working Plan 1978- 2000 and Sahu' s Working Plan 1984-2000

S.No. Working Circle	*Working plan*	*Silvicultural system*	*Forest to which applied*
1. Selection-cum improvement	Goyal's Plan	Selection and improvement felling	Dominantly matured age crop
2. Conversion	Goyal's Plan	Conversion	Dominantly matured age crop standing in the site quality III area
3. Coppice with Reserve	Goyal's Plan	Coppice	Young aged coppice crop near habitation
4. Bamboo (overlapping)	Goyal's Plan	Bamboo felling with four years cutting cycle	Bamboo forests
5. Improvement	Sahu's plan	Improvement felling	Dominantly young
6. Soil and Water conservation	Sahu's plan	Artificial regeneration	Erosion prone areas
7. Fuelwood Plantation	Sahu's plan	Artificial regeneration	Site quality V areas near habitation

The story of management of protected forests is somewhat different from reserved forests. Before 1989, protected forests were either categorised as Zamindari Forest or Private Forests. There was absolutely no ground restriction on felling of trees on such forests. Villagers used to fell trees of their choice to meet their personal requirement or Nistar need. The felling remained uncontrolled. But, because of low anthropogenic pressure on forests, the interior forests were not severely affected. Only, the forests near habitations were degraded. In the year 1889, some rules under Land Revenue Code-1889 were formed. This was required to be enforced by Zamindars. But, Zamindars failed to enforce it. Therefore, it also left no impact on the degradation process of the forests situated near villages. Rather this period witnessed more felling near the villages. Because of this, an attempt was made by local feudal authorities to enforce Wazib-Ul-Urge – a regulatory mechanism to control felling. But it also failed to fully regulate the illegal fellings (Table 3.4).

In spite of host of initiatives taken by the authority, no mechanism was fully effective. In the year 1951, these forests were acquired by provincial government of CP & Berar under Abolition of Property Rights Act. These forests were transferred by revenue department to forest department in the year 1953. Then forests were declared Protected Forests under Section 4 of IFA-27. Forest department took initiatives to conduct survey and demarcation of such forests and to prepare working scheme/working plan and manage these forests on scientific lines.

Results of Management in the Last Twenty Years

Forest Management planning is a continuous process. It needs to be monitored and evaluated periodically. If any corrective measures are required to improve upon management practices, than they ought to be taken up in opportune time. Otherwise, the impact of management practices on the forest will be inimical. However, for undertaking corrective measures, it is necessary to evaluate the management plan on a periodic level and identify the micro and macro areas where corrective measures are necessary. Evaluation of the plan should take place at an interval of at least five years or ten years.

The impact of management practices on the forest of the study area during the currency of the last working plans can be evaluated at two levels. The first level can be evaluation at management level, *i.e.* working circle level. The second level can be at ecosystem level, *i.e.* evaluation with respect to different facets of forest ecosystem.

1. Management Level Appraisal

The impact of management practices on forests can be evaluated working circle-wise on the basis of specific management objectives determined for that circle in the last working plan. The impact on forests of different working circles are coalesced to develop a pattern which will indicate the impact of management on forest in totality.

2. Ecosystem Level Appraisal

The impact of management practices on the forests can be studied from an ecological point of view. The basis of evaluation and impact analysis is changes seen in ecosystem structure and functions. This is done by studying the changes occurred with reference to the base line data. The parameter which can be used are forest cover, forest areas, density-wise areas, forest composition, tree density, stem distribution, regeneration status, growing stock, bio-diversity, productivity etc.

Management Level Appraisal

Impact evaluation on the basis of changes found in the forests of different working circles helps in understanding the extent to which the objective of management has been met. In the study area, there were two working plans in operation. In both the plans, the working circles were :

1. Selection-cum-Improvement W/C
2. Conversion W/C
3. Plantation W/C
4. Coppice with Reserve W/C
5. Soil Conservation-cum-Improvement W/C
6. Bamboo (Overlapping) W/C

Selection-cum-Improvement Working Circle

Selection-cum-Improvement (SCI) was a very important working circle. It covered an area of 96393 hectares and constituted about 39 per cent of the study area. The treatment extended to areas of Conversion and Plantation working circle also because of moratorium imposed on clear felling by the state government in the year 1984. The area of operation under this circle spread almost all ranges of the study area. The specific objective of management of this circle was to augment the regeneration status and also to improve the growing stock. The secondary objective was to encourage the spread of the teak forests by promoting preferential treatment to teak advance growth at the expanse of mixed crop. Against these objectives, the teak area as well as its proportion in the growing stock improved. There is also evidence that regeneration status in teak forests has improved.

Felling of trees is an important function of production forestry. It improves the light availability on the floor of the forests. It indirectly promotes not only seed germination but also stand development. Therefore, every working plan for production forestry emphasizes on felling of matured trees and improvement of crop. In the study area, the degree of felling in a unit area was analysed by examining the felling data of the coupes in the last plan period (Table 3.5). The number of trees felled per hectares was about 62. Out of it, the trees having exploitation girth ranged from 6 to 8. The rest were young and middle aged trees. These trees were prematurely removed under thinning and improvement treatment practices (Table 3.6). From the point of timber and fuelwood production, the yield per hectares was 3.31 cmt and 12.619 cmt respectively. This constituted a ratio of 1:4 (Table 3.7). This looks very high. It may be because of inefficient logging and imperfect market condition prevailing in the study area in that point of time.

Table 3.5 : Felling Intensity Per Hectares

S.No.	Working circle	No of trees in different girth classes					
		21 to 45 (cm)	46 to 60 (cm)	61 to 90 (cm)	91 to 120(cm)	121 to above (cm)	Total
1	SCI/Conversion to Uniform /Plantation	19.373	14.056	16.559	6.541	5.718	62.247
2	CWR	56.056	18.977	16.376	7.867	4.455	103.731

Preference was given to tree species such as Teak, Saja, Dhawda, Haldu and Bija. Mixed species other than the valuable species were not given preferential treatment so they were felled. They constituted about 15 per cent of the total trees felled in a unit area. As a result, the proportion of prefered timber species in the forest composition improved considerably.

Table 3.6 : Species-wise Felling Intensity

S. No.	Working circle	No of trees felled per hectare					
		Teak	Saja/Salia	Bija/Tinsa/ sisam	Haldu/Mundi/ Tendu/Dhawda	Other	Total
1	SCI / Conversion to Uniform/ Plantation	2.124	4.862	2.289	4.666	48.306	62.247
2	CWR	3.239	10.439	2.855	36.685	50.313	103.74

Table 3.7: Production trend in Different Working Circles

S. No.	Working Circle	Production Per Hectare		Total Production (in cmt)	Timber and Fuel wood Proportion
		Timber / Poles	Fuel Wood		
1.	SCI	3.331	12.619	15.950	1: 3.79
2.	Conversion to Uniform	4.546	10.796	15.342	1: 2.37
3.	Plantation (Teak)	9.080	43.129	52.209	1: 3.92
4.	Weighted Average (SCI/ Conversion/ Plantation)	4.03	15.79	19.82	1: 4.05
5	Coppice with Reserve	4.603	18.644	23.247	1: 4.05

Regeneration status in the forests is not encouraging. The incidences of grazing and fire in the forest are high. This might have been one of the reasons for soil compaction in the forests. Declining bio-diversity in forests might have led to decrease in population of pollinators. A major part of the soil in the study area is black cotton soil, which does not promote good tree regeneration. All these factors have slowed the process of natural regeneration in the forests. But, long patches of open forests are not seen. In some protected area, artificial plantations were also raised. In the last plan period, an area of 2442.77 hectares was artificially regenerated.

Constitution of the Selection-cum-Improvement (SCI) working circle aimed at increasing production of forest produces such as timber, poles and fuelwood. Annual timber production in the study area varied from 15000 to 30000 cmt. Along with fuelwood, it came to about 50000 to 80000 cmt per year. Sixty per cent of this production came from area of SCI working circle. The annual productivity varied from 0.3 cmt to 0.7 cmt per hectares (Table 3.8).

Table 3.8 : Plantation in Different WP Periods

S. No.	*Working Plan*	*Teak*	*Mixed*	*Bamboo*	*Total*
1.	Previous to Current WP	2517.01	120.00	135.00	2772.01
2.	During Goyal & Sahu WP	3651.50	13327.65	2349.58	19328.73
	Total	6168.51	13447.65	2484.58	22100.74

Conversion Working Circle

The area under this working circle was 11733.39 hectares. But, due to State Government's ban on clear felling of forests, the working circle could not be operational. Therefore, the prescriptions of SCI were applied to this circle. Consequently, the results of this also tallied more or less with that of SCI working circle. Production from such forest remained around 15 cmt per hectares, 1 cmt less than the SCI working circle. Timber and fuelwood proportion remained at 1 : 2.37, *i.e.*, relatively better than other working circles. Preference for the valuable species was given. As a result, density of teak, saja, mode, salia in the forest composition increased. Area under this circle contained more of teak forests. The percentage of teak trees increased at the cost of mixed crops. This was the objective of SCI also. The status of regeneration was better than that of SCI. The reason might be that forests of this circle are of better site qualities, (more of III site quality) and are located on well drained soils. Besides, low anthropogenic pressure on such forests could also be one of the reasons for improved status of regeneration.

Plantation Working Circle

The area allotted to plantation working circle was 10144.06 hectares. This constituted about five per cent of the total plan area. The prescriptions were not implemented because of the State Government's ban on clear felling of forests. As a result, prescriptions of Selection-cum-Improvement were implemented in the area of this working circle. However, artificial regeneration was undertaken in gaps available in the forests up to the year 1996. An area of 8619.66 hectares was treated as per the prescription of Selection-cum-Improvement.This constituted about 68.8 per cent of the working circle. Gap planting under artificial regeneration strategy was taken up in an area of 2351.96 hectares. Out of it, teak, bamboo and mixed plantation constituted about 1648.88 hectares, 328.08 hectares and 375.00 hectares respe-ctively. Most of the plantation areas are found in Manpur and Chowki ranges. These areas are mostly plane and have sandy loamy soil.

Plantation constituted an important function of management under all working circles. Wherever, areas suitable for artificial regeneration were found, plantation was carried. Therefore, plantation activities continued not only in plantation working circle but also in other working circle areas too. The study area has 22101 hectares of various plantation distributed all over the ranges of the study area. Statistics regarding plantation in different working plan periods and working circle areas are presented in Tables 3.9, 3.10 and 3.11. Mixed plantation has been raised in all working circle area. But, these plantations have failed.

Table 3.9 : Plantation in Different Working Circles in the Study Area Excluding Gandai Range (up to Goyal WP year 1995)

S. No.	*Working Circle*	*Teak*	*Bamboo*	*Mixed*	*Total*
1.	Selection-cum-Improvement	356.72	429.00	1657.05	2442.77
2.	Soil and Water Conversion	693.40	844.00	5584.40	7121.80
3.	CWR	0.50	130.00	3362.00	3492.50
4.	Conversion to Uniform	308.00	70.00	158.20	536.20
5.	Plantation	1648.88	328.08	375.00	2351.96
	Total	3007.50	1801.08	11136.65	15945.23

Table 3.10 : Plantation in Different Working Circles of Gandai Range (up to Sahu WP year 1996)

S. No.	Working Circle	Teak	Bamboo	Mixed	Total
1	Improvement Felling	–	148.00	385.00	533.00
2	Soil & Water Conservation	320.00	200.00	270.00	790.00
3	Teak Plantation	74.00	–	230.00	304.00
4	Energy Plantation	250.00	200.50	1306.00	1756.50
	Total	644.00	548.50	2191.00	3383.50

Table 3.11 : Plantation in Panabaras Project Area of Forest Development Corporation (till year 2001)

S.No.	Species	Area under Plantation in hectares
1.	Teak	17789
2.	Bamboo	10337
3.	Miscellaneous	28
	Total	28154

Note : The Panabaras Project area extends to 41404 hectares.

An analysis of the timber production consequent to felling in working circle areas indicates that the forest yielded highest per hectares productions. The production per hectares in case of timber was 9.08 cmt and in case of fuelwood it was 43.129 cmt. The forest of this working circle mostly constituted of III quality and mature crop. Therefore, productivity of this working circle was very high in comparison to other circles.

The impact of forest management over the years is perceptible. There is an improvement in crop condition and proportion of timber trees in the natural forests. However, regeneration is not satisfactory. Since, the forest land of this circle were very productive, it attracted the eyes of encroachers. The encroachers were local tribals. The farmers who had forest land adjoining to their land, gradually invaded the fertile forest land. Over the years, anthropogenic pressure on the forests increased manifold. This has a telling effect on forest regeneration.

Coppice-with-Reserve Working Circle

The extent of this working circle is 28453.490 hectares. It accounted for 13 per cent of the plan area. The felling cycle was 40 years. The production profile of this working circle is very conspicuous. It indicates a very high rate of harvesting. What is important to note is that the timber outturn was only 4.603 cmt per hectares and the fuelwood was 18.644 cmt, taking the total outturn from a hectare area of forests to 23.247 cmt. Since, the crop was being maintained basically for pole crops, the outturn was very obvious. More than 75 per cent of this outturn came from crop having the girth class of 21 to 60 (Table 3.5). Till the year 1996, a total area of 11229.40 hectares was treated. The site quality of this circle varied from VA to V B. The crop of this circle basically ranged between the girth class of 21 to 45 cm., *i.e.* basically young aged.

Trees of saja, salia, dhawda, karra, senha, mundi were felled. Species such as teak, bija and tinsa were given preference at the cost of other mixed species. Because of intensive felling of vigorous coppicing species like Karra and Bhirra Lendia, these crops increased their density. In the forests ranges like Bhagnadi, Rajnandgaon and Dongargarh, the population of bhira increased perceptibly. A decline in the population of bija was also observed. The natural regeneration has become a problem primarily because of incidences of illicit felling near villages and town areas. Coppice regeneration in this forest is satisfactory but, seedling regeneration of teak, bija, saja, mundi and dhawda is poor. Till the year 1996, 3992.5 hectares of gaps were planted. Out of it, mixed plantation constituted about 3362.0 hectares. The rest were either teak or bamboo plantation.

Soil Conservation-cum-Improvement Working Circle

The area allotted to this working circle was 20226.54 hectares. This constituted about 9 per cent of the plan area. Till the year 1996, 7121.80 hectares of area were either planted or treated with soil conservation works. Out of it, teak was planted in an area of 693.40 hectares, bamboo in 844.00 hectares and mixed plantation in 5504.40 hectares. The result of most of the teak and bamboo plantations are very good. But mixed plantations conspicuously failed everywhere. Since then plantations were taken after undertaking cut-back operation in the forest and so ample coppice growth emerged in such areas. Resultantly, the stocking in the forests improved.

Some of the compartments under this working circles were taken up for rehabilitation under Joint Forest Management (JFM) programme. Most notably, Gatapar and Jantar in Rajnandgaon range, Sangli and Maldongri of Chowki range. JFM committees undertook rehabilitation and protection work. Today, the forests, which were once exposed and highly under-stocked blossomed into a patch of good dense forests. This was possible because of the people's involvement in protection and development of the forests. More details are presented in separate chapter focusing on participatory forest management.

Baring few patches where local people took immense interest in developing the forests, most of the area of this working circle did not recover to the point where it was in the 1960s or 1970s. Such areas are still under heavy anthropogenic pressures. Incidences of fire, illicit felling, and encroachment are occurring frequently. As a result, most of these forests have undergone serious degradation, both in soil and vegetational conditions.

Bamboo (Overlapping) Working Circle

The working circle extends to an area of 65419 hectares. This accounts 27 per cent of the total plan area. Bamboo felling had been carried out in accordance with the bamboo felling rules prescribed by the State Government. But, it has been observed that there is a distinct decline in the productivity of Nistar coupes.

Leaving aside the aspect of declining productivity in Nistar felling series, the overall productivity in dense bamboo forest is very good. The productivity is very high in bamboo site quality I and II areas. These types of area account for 85 per cent of the total bamboo forests. The productivity in such area ranges from 0.8 to 1 NT per hectares (Table 3.12).

Table 3.12 : Bamboo Productivity Trend: Period 1990-1997

S.No.	*Year*	*Bamboo Production (NT)*		*Annual bamboo productivity per hectares in NT*
		Commercial	*Industrial*	
1.	1990-91	0.27	0.62	0.89
2.	1991-92	0.28	0.53	0.81
3.	1992-93	0.45	0.49	0.94
4.	1993-94	0.33	0.47	0.80
5.	1994-95	0.30	0.55	0.85
6.	1995-96	0.18	0.38	0.56
7.	1996-97	0.30	0.53	0.83

Note : Total exploitation area of bamboo was 81187 hectares.
Source : Working Plan of Rajnandgaon:1999.

Most of such bamboo forests where production declined in the last plan period were Nistar bamboo coupe areas. They were situated near habitation. They received uncontrolled biotic pressure in the form of illicit felling, fire and grazing. As a result, all of the Nistar felling series manifested a distinct decline in bamboo productivity (Table 3.13).

Table 3.13 : Bamboo Forests Degraded during Goyal Plan

S. No.	*Range Name*	*Felling Series*	*Area of Compartment*	*Compartments*	*No. of Compt.*
1.	Manpur	Muchur	3031.09	626, 627, 632, 633, 634, 637, 699, 700	8
		Dorde	1172.76	701, 702, 703, 704, 706	5
		Naharmeta	1425.71	719, 720, 724, 725, 726, 727	6
2.	Dongargarh	Lingangarh	1874.23	170, 171, 213, 214, 217, 241, 242, 245, 246	9
3.	Rajnandgaon	Markampur	1301.05	337, 338, 339, 344, 345, 355	6
		Total	8804.84		

Note : All such forests have witnessed degradation, as the productivity decreased from 0.2 Notional Tonne (NT) per hectares.
Source : Working Plan of Rajnandgaon:1999.

The areas which suffered a drastic decline in density constituted about 8805 hectares. It constituted about 21 per cent of the productive bamboo area. These coupes were situated in ranges like Rajnandgaon, Chowki, Dongargarh and Manpur. These felling series used to give an yield of 1 NT or more per hectares in the 1970s, but over years the productivity declined to 0.2 NT per hectares. What is more disconcerting is the fact that among these Nistar felling series, 15 compartments which had good dense bamboo forests, completely degraded and fell into the category of degraded bamboo forests. This retrogression took place in a time span about 20 years. The process of degradation of bamboo forests in Manpur, Panabaras and Chowki range is still continuing. The same threats hang over the bamboo forests of Khairagarh and Dongargarh also. Extraction of bamboo from such forests are alarming and mostly beyond the forest's capacity to recover. Generally, it was observed that the bamboo felling rules were not observed strictly. The rule regarding retention of minimum bamboo clumps in each clump has not been followed properly. In other words, the harvesting is more than the annual increment of the bamboo growing stock.

In Gandai range, bamboo forest extends to an area of 14,147 hectares. Here, the site quality and density of bamboo are low. Therefore, these forests were never treated as productive bamboo forests. In an analysis, it was found that out of nine bamboo felling series, coupes of five felling series suffered from heavy degradation. As a result, the productivity severely declined (Table 3.14).

Table 3.14 : Productivity Analysis of Bamboo forest of Gandai Range

S. No.	*Felling circle*	*Area in (hectares)*	*No of compartments*	*Productivity class*
1.	Narori	1312.32	9	+++
2.	Lamara	1606.66	7	+++
3.	Koparo	1997.26	8	+++
4.	Dhapla	1948.21	8	+++
5.	Botesur	1232.02	5	++
6.	Balrampur	2482.02	8	++
7.	Topo	844.01	6	+
8.	Piparahi	1485.77	8	+
9.	Agarawara	1288.59	8	+
	Total	14,196.86		

Source : Working Plan of Rajnandgaon, 1999.

A general decline in terms of extent, quality and growing stock of bamboo forest has been witnessed in all parts of the study area. Resultantly, productivity of the bamboo forests decreased. The bamboo inventory study shows that there is an asynchronous distribution of different ages of bamboo clumps. Since bamboos are worked on a four year working circle (rotation) for sustained yield management, the distribution of different ages of bamboo clumps should be : (1) 25 per cent of 1st year bamboo (Karla), (2) 25 per cent of 2nd year bamboo (Mahila) and (3) 50 per cent of the 3rd year and 4th year bamboos (Pakiya). Against this norm, the study area presents an asynchronous pattern of distribution. The Karla constitutes 16 per cent, *i.e.* 9 per cent below the ideal mark, Mahila 28 per cent and Pakiya 56 per cent, *i.e.* 3 per cent and 6 per cent higher than the required mark. This means the bamboo recruitment is far below the sustainable point and harvest is also 6 per cent below the required mark. Low yield is generally attributed to presence of more dry and damaged bamboo in the clumps. Both these points indicate the fact that fire incidences in the forests not only retard the recruitment of news bamboo culms but also increase the proportion of dry bamboos in the clumps.

Ecosystem Level Appraisal

In the year 1980, Forest Survey of India (FSI), Dehradun, conducted a forest inventory of Rajnandgaon, Kawardha and Durg districts. They completed the survey work in the month of March 1981 and submitted a report on 1985. During this period, the last working plan was also prepared. So, the inventory prepared by Forest Survey of India was taken as " Base Line Data". Again in the year 1997, a fresh inventory of Forest Resource Survey under the aegis of working plan was carried out in the study area by the author. The time span dividing the two inventories is about 17 years. A time span of 17 years or 20 years is an appropriate period for an ecosystem level appraisal. Therefore, an in-depth analysis was conducted to find out the changes that have taken place in forests ecosystem over the period of seventeen years.

Forest Survey of India divided the forests into two broad strata *i.e.* teak forest and the mixed forests. This apart, an integrated inventory of the forest of the Rajnandgaon–Durg catchment was prepared. Therefore, in order to compare with the base line data, the results of the present resource survey were also analysed by taking similar critical parameters. Forest Survey of India undertook measuring trees having girth size about 30 cm, where as in the present resource survey, trees having girth above 21 cm were measured. But, in order to make the comparison possible, the girth class of 21 to 30 cm in the present survey was not taken into consideration. In addition to this, Forest Survey of India, calculated the growing stock by using local volume table developed by FSI itself. Therefore, during the working plan exercise, the growing stock was also calculated by the author with the help of using the same local volume table developed by FSI.

The parameters for appraisal at ecosystem level are as follows :

1. Changes in forest area.
2. Changes in forest types and succession.
3. Changes in forest composition.
4. Changes in site quality.
5. Changes in crown density.
6. Changes in crop volume.
7. Changes in growing stock: (1) Teak, (2) Mixed, and (3) Bamboo.
8. Changes in status of regeneration.
9. Changes in productivity.
10. Changes in plantation area.

Changes in Forest Area

Shrinking of forest is a conspicuous world phenomenon. The society is concerned about it. Therefore, it is worth considering weather there is any shrinkage of forest area or not in the study area. In addition to it, since this being a micro study at landscape level, it is pertinent to examine the change in areas of forest ecosystems in the division. Broadly from the point of management, we can carve out three ecosystems: (1) teak forests, (2) mixed forests with bamboo, and (3) mixed forests without bamboo. An account of change in forest area and composition is presented in Table 3.15. The analysis brings forth the following points:

Table 3.15 : Change in Forest Area and Composition

S.No.	*Composition*	*Year 1978*		*Year 1997*	
		Area (hect.)	*% of Forest Area*	*Area (hect.)*	*% of Forest Area*
1.	Teak Forest				
	Dense	18860.24	8.36	17031.713	7.53
	Degraded	–	–	366.737	0.16
	Total	18860.24	8.36	17398.450	7.69
2.	Mixed Forest				
	Dense	175133.31	77.62	161755.524	71.53
	Degraded	26599.657	11.79	24311.653	10.75
	Total	201732.96	89.40	186067.177	82.28
3.	Blank Forest and Plantation	5042.108	2.23	18011.578	7.97
4.	Encroachment	–	–	4451.540	1.97
5.	Nala, River bed, Inundated area, other	–	–	200.743	0.09
6.	Bamboo Forest Area	65419.24	28.99	61499.291	27.20
	Total	225635.315	100.00	226129.488	100.00

Source : Working Plan of Rajnandgaon, 1999.

There is a qualitative and quantitative change in teak forest over the period of 20 years. In the year 1978, all the teak forests were of dense category that constituted 18860 hectares where as in the year

1997 the same area shrunk to 17032 hectares. This apart, 367 hectares of dense teak forest got degraded. What is more significant is the fact that an area of 1462 hectares of teak forest gradually changed into mixed forests, primarily because of diminution of teak tree population.

There is also a qualitative as well as quantitative change with respect to mixed forests. In the year 1978 the extent of dense mixed forests was 201733 hectares. But in the year 1997, the extent of such dense forests declined to 186067 hectares depleting about 15666 hectares. Further discounting 367 hectares of teak forest, which got converted into mixed forests, the net depletion remained at 15299 hectares.

There is a loss of 15299 hectares of forests within a period of about 20 years. Therefore, the rate of forest depletion comes to about 765 hectares annually. Since the process of deforestation begins with degradation and ends with complete depletion, it can be reasonably concluded that at least, if not more, an equal area of forests are also degraded every year. It means every year, about 800 hectares of forests are degraded and an area of equal size is depleted.

The encroachment area enlarged enormously over the period of 20 years. About 4451.54 hectares have been found to be encroached. If we analyze it, it is found that the annual rate of deforestation on account of encroachment is 223 hectares. A major chunk of encroachment has been found in the range Manpur, the southern tip of the study area. This range comes in the Mohla tahsil. Here the soil is fertile and deep, so highly suitable for agriculture. At present the population density in this tahsil is 81, which is the minimum in any range. But majority of population in this range is tribal, and they are poor. What is more disturbing is the fact that the population growth rate in the last decade, *i.e.* 1981 to 1991 was 24.96 per cent, which was highest in any tahsil.

The total plantation in the study area till the year 1978 was about 5042 hectares. But over the past period of 20 years, the total plantation rose to 18011 hectares. Thus an area of 12970 hectares plantation was raised in the period of twenty years. The annual rate of plantation comes to about 650 hectares.

The rate of planting (artificial regeneration) does not match the rate of degradation and depletion. The rate of depletion is about 765 and the rate of degradation is at least of same order. As a result, the net degraded forest in the study area still remained at 24679 hectares in spite of efforts to regenerate.

The proportion of dense and degraded forest in the beginning of the last plan period was 6 : 1. Now in the year 1997, it has increased to 5 : 1. This indicates that the process of degradation has seriously set in.

The most striking results come from the analysis of bamboo forests. There are three types of changes that have been witnessed. First is depletion of bamboo forests, second is degradation of bamboo forests and third is decline in productivity of remaining bamboo forests. We have already discussed about decline in productivity of bamboo forests in previous paragraph of this chapter. Here we shall be discussing about depletion and degradation of bamboo forests. In the year 1978, the extent of bamboo forest was 65419 hectares. But in the year 1997, the bamboo area remained at 61499 hectares. That means there was a net depletion of 3920 hectares. By this account, the annual rate of depletion of bamboo forests comes to about 200 hectares. The total degraded bamboo forests in the year 1997 stood at 8805 hectares. Putting both the phenomena of degradation and depletion in a cyclic phase, we can conclude that annual degradation and depletion of bamboo forests is taking place at a pace of 440 hectares and 200 hectares respectively in two consequent cycles. In other words, the annual rate depletion of bamboo forest ranges from 10 to 13 per cent.

Degradation of bamboo forests is also found along a resource continuum. The site quality, density and terrain act as an integrated framework of continuum. The anthropogenic pressure as well as current management regime play as a set of driving forces, which lead the degradation process in the bamboo forests along the continuum. In case of increasing pressure on forests, the bamboo area having site quality III and low density become the first victim. Then the area having site quality II and low density gets degraded. Likewise the process of degradation moves in the study area.

During the last twenty years, an area of 2350 hectares were rehabilitated with artificial regeneration of bamboo. This accounts about 130 hectares in every year. This does not match the rate of either depletion or degradation. There was no planned rehabilitation programme in the last plan as such. This apart, the degraded bamboo forest remained confined mostly in ranges like Manpur, Dongargarh and Rajnandgaon. In Manpur range, the probable primary cause for the depletion appears to be more of ecological than anthropogenic in nature. This range is more of undulating in nature, unlike that of other hill ranges where bamboo occurs. In this range, on account of forces of succession, the dry deciduous mixed forest with bamboo is converted into dry deciduous mixed forest without bamboo. This apart, increasing incidences of fire also played a role in facilitating the process of degradation.

Changes in Forest Types and Succession

Succession is a dynamic ecological process. It, at one point, focuses on endpoints such as climatic climax, edaphic climax or successive stages or path ways. Recent research in succession has helped to demystify the process of succession. In the classical model of Clements (1916), succession was a temporal sequence of plant communities, each of which changed the local environment and facilitated invasion by the next community. The process that drove this model included disturbance, migration, acesis or establishment, biotic reaction or environmental change, and competition.

The endpoint of these processes was stable climax vegetation that was the same for merely all sites in climatic region. In the study area, four important forest types are found. Applying the Clement's theory, the succession status in all the types are edaphic climax (Table 3.16). Over the year, other theories of succession have come. Picket *et al.* (1995) has proposed a hierarchical framework for vegetation dynamics that includes disturbance and biotic processes. Particularly important biotic processes behind succession are colonization, competition growth rate and longevity/mortality. Instead of viewing succession from only ontological perspective (*i.e.* emphasis on processes and casual mechanisms), a phenomenological perspective (*i.e.* a description of patterns, pathways and probabilities) are useful in management application. Therefore, the concept of climax as a stable endpoint is misleading. It is uncommon for vegetation or ecosystem to reach equilibrium point or condition of no change, which requires a perfect balance between opposing forces or disturbance and biotic development (Picket and White, 1985). Vegetation dynamics can be

Table 3.16 : Forest Types and their Succession Status

S.No.	*Title*	*Forest types*	*Succession status*
1.	Dry teak forests	5 A/c	Edaphic climax
2.	Northern Dry mixed Deciduous forests	5 B/C2	Edaphic climax
3.	Boswallia forests	5/E2	Edaphic climax
4.	Dry Bamboo Brakes	5/E9	Edaphic climax

characterized as shifting mosaic or patches or different ages and developmental stages. Patches of vegetation changes like a kaleidoscope. Spatial aspects of this concepts provide a framework for how spatial characteristics such as patch size and juxtaposition of patches affects overall system behaviour. The importance of canopy gap dynamics to forest dynamics is a case in point. It is said that gap creation rates in forest range between 0.2 per cent to 2 per cent of a stand each year.

Out of these four major forest types in the study area, Boswallia forests have shrunk considerably over the years. The possible reason for the dwindling population is lack of regeneration. This species generally comes from root suckers. Because of fire and incompatible changes in the forest floor, regeneration has been severely affected. Boswallia forests are formed in hilly parts of Dongargarh and Khairagarh. Bamboo forests adjoining to habitations and in unproductive sites have shrunk. Now, these degraded patches are expanding. Bamboo brakes also suffer from ecological transformation. Dry mixed deciduous forests with bamboo is gradually converted into dry mixed deciduous forests without bamboo.

Forests, whether edaphic climax, pre-climax stages or any other stages are liable to be degraded to varying degrees through anthropogenic influences, notably uncontrolled fire, excessive grazing, lopping for fuel or illicit felling. Such action is often selective and concentrated on the species most valued for several purposes, and not merely to the extent of virtual elimination of the desirable species. The seedling stages are particularly exposed to extermination. The process is also influenced by verifying process of resistance of individual species to those treatment or changes enabling the hardiest at least to survive while more susceptible are exterminated.

Changes in Forest Composition

Management practices in the last plan period left an indelible impact on forest composition. Population of certain species declined and certain species increased. Possibly, this happened primarily due to management practices adopted in the last plan. One of the important management objectives of the last plans was to increase the population of valuable species. Valuable species mean species which have high timber value. Those species were teak, bija, saja, tinsa. Out of these teak was given utmost priority.

An analysis of impact on teak and mixed forests has been done. The following change has been observed in forest composition :

Teak Forests

1. Considered from a point of entirety, the density of tree species as such almost remained unchanged. There is an increase of only one per cent in population size. A change of one or two per cent in tree density is not significant looking at the broad spectrum of forest composition.
2. Population of teak, and saja and other species increased. They are the important timber species in teak forests. Teak increased by 186 per cent, saja by 67 per cent, mode by 73 per cent. Apart from this, species like Bhirra, Garari, Saliya increased by 16 per cent, 45 per cent and 27 per cent respectively. This has resulted because species like Bhirra is a pioneer species and Garari is a strong coppicer. It appears, the management practices did not target to increase their population, but it did increase because of their aggressive ecological and silvicultural characteristics (Table 3.17).
3. There is a decline in the population of few species such as tendu, dhawda, tinsa, bija, lendia and other species by about 47 per cent, 32 per cent, 46 per cent, 2 per cent, 38 per cent and 64 per cent respectively. The management aimed to decrease the population of lendia and many other non-timber species. The population of lendia declined but along with other important species. Sisham species also declined drastically (Table 3.17).

Mixed Forests

1. There is no big change in the tree density. There is a marginal change of only 2.2 per cent. Therefore, looking at this, we can conclude that there is no change.
2. There is an increase in the population of teak by 120 per cent. This is a very significant change in the mixed forest. Though the management targeted this in its plan, but not to this scale. Looking at the over all stability of the tree population, increase in teak population have lead to a decrease of some other species (Table 3.17).
3. There is an increase in population of saja, mode, bija, lendia and bhirra species by 32 per cent, 14 per cent, per cent, 60 per cent and 37 per cent respectively (Table 3.17).
4. There is a decrease in population of tendu, tinsa, garari and salia species by 48 per cent, 37 per cent, 15 per cent and 27

per cent. Out of these the management targeted to decrease the population of garari and it has borne fruit. The reason for declining population of tendu and salia is basically destructive harvesting practice as well as over exploitation found in the study area. Mortality of tendu was also reported because of pernicious practice of debranching the tree for tendu leave collection in some pockets (Table 3.17).

Table 3.17 : Change in Tree Density: Teak and Mixed Forests (Girth Class above 30 cm)

(Trees per hectare)

S. No.	*Species*	*Teak Forest*			*Mixed Forests*		
		1980 Resource Survey	*1997 Resource Survey*	*Change in %*	*1980 Resource Survey*	*1997 Resource Survey*	*Change in %*
1.	Teak	29.242	83.542	+ 186	5.078	11.200	+ 120
2.	Saja	18.005	30.00	+ 67	25.834	34.035	+ 32
3.	Dhawda	16.742	11.459	- 32	20.099	20.303	+ 1
4.	Tendu	12.986	6.875	- 47	15.696	8.106	- 48
5.	Tinsa	10.492	5.626	- 46	10.849	6.786	- 37
6.	Bija	7.221	7.083	- 2	7.457	7.715	+ 3
7.	Mode	8.646	14.999	+ 73	11.239	15.265	+ 14
8.	Lendia	8.434	5.208	- 38	12.163	19.472	+ 60
9.	Bhirra	6.347	7.334	+ 16	8.132	11.118	+ 37
10.	Garari	17.034	24.658	+ 45	28.924	24.588	- 15
11.	Saliya	7.286	9.241	+ 27	12.802	9.282	- 27
12.	Other	94.352	33.97	- 64	103.871	88.503	- 15
	Total	236.787	239.995	+ 1	262.144	256. 373	+ 2.2

Source : Resource Survey (Working Plan of Rajnandgaon:1999).

Forest degradation has also impacted severely on the forest composition. With the onslaught of anthropogenic factors, degradation has set in many forest patches. It has not only degraded the forests but also affected the forest composition. There are certain species whose population has increased in such degraded and open forests. Examples are bhirra, palas, and to some extent bahera. The population of bhirra and palas has increased at a rapid pace. This apart, there are few species whose population is decreasing due to degradation. Examples are bija, amla, karra, dhawda, tinsa, tendu, achar mode and saja. The reason for such change is perhaps increasing dryness of the area. This might have been arisen due to rise in fire intensity as well as frequency in forest areas, especially fringes. Along with fire, the increased fragmentation of the forests has also impacted on forest composition very adversely.

Changes in Site Qualities

In forestry site quality does not change normally in a short span of time. It may change in a span of 500 years or more. But, the method of measurement may change over time. In fact, the method of measurement has improved over the years. More modern instruments are being used to measure the top height of a crop. This has improved the precision of height measurement. However only very young teak plantations and forest crop have transited to middle aged crop, which would have changed the estimation of site quality. It is because site quality is measured on the basis of maximum tree height attained in that area. Looking at the degree of change in areas under different site qualities in the study area, one can safely assume that the site quality estimation is fairly accurate and no major change is perceptible (Table 3.18).

Table 3.18 : Change in Site Quality

S. No.	*M.P. Site Quality Class*	*Percentage of area during the last plan*	*Percentage of area in the current plan*
1.	III	31.44	27.48
2.	IV A	46.91	47.35
3.	IV B	20.05	22.30
4.	V A	0.60	2.87
	Total	100.00	100.00

Source : Working Plan of Rajnandgaon, 1999.

Change in Crown Density

From the point of crown density, 89 per cent of the total forests are having a crown density more than 0.5 and 11 per cent of the area has density below 0.4. The crown density of the crop in dense forest category ranges from 0.5 to 0.7 and in some sites it ranges up to 0.8. In some ranges like Chowki, Rajnandgaon and Gandai, there is a perceptible decline in crown density over the period of about 20 years.

Change in Crop Volume

Volume in teak forests in the year 1997 is 63 cmt per hectare. This was about 58 cmt per hectares in the year 1980. Likewise, volume in mixed forests in the year 1997 is 64 cmt per hectares. In the year 1980, it was 63 cmt per hectares. Therefore, with respect to volume, there is no substantial change in the crop volume in the span of about 20 years.

A change has been noticed in the growing stock of certain tree species. In the year 1980, teak comprised of 15 per cent of the total growing stock in teak forests. But in the year 1997, it increased to 27 per cent of the total growing stock of forests. Likewise, in the year 1980, the mixed crop comprised about 86 per cent of the total growing stock of the teak forests. In the year 1997, it decreased to 73 per cent of the total growing stock, registering a fall of 8 per cent. It indicates that, the objective of management, which aimed at improving the valuable timber species, was fulfilled. But, this happened at the cost of diversity of miscellaneous species. Though, the extent of teak forests in the study area is not so large but, it has sizable presence (Tables 3.19 and 3.20).

Table 3.19 : Change in Volume of Teak Forests (Volume per hectares in cmt)

S.No. Species	*1980 Resource Survey*	*1997 Resource Survey*	*Change in %*
1. Teak	8.414	17.111	+ 103
2. Saja	6.015	8.79	+ 46
3. Bija	2.788	3.621	+ 30
4. Tinsa	1.072	1.887	+ 76
5. Tendu	2.545	1.383	- 46
6. Dhawda	4.150	4.283	+ 3
7. Lendia	0.885	0.811	- 8
8. Salai	4.617	6.281	+ 48
9. Garari	1.225	0.098	- 92
10. Other	26.524	17.991	- 32
Total	58.235	62.796	- 8

Source : Resource Survey (Working Plan of Rajnandgaon, 1999).

Table 3.20 : Change in Volume of Mixed Forests (Volume per hectares in cmt)

S. No. Species	*1980 Resource Survey*	*1997 Resource Survey*	*Change in %*
1. Teak	1.699	0.909	- 46
2. Saja	8.128	11.262	+ 39
3. Bija	3.071	4.611	+ 50
4. Tinsa	1.362	2.298	+ 69
5. Tendu	3.819	2.677	- 30
6. Dhawda	5.503	5.992	+ 9
7. Lendia	1.437	1.977	+ 38
8. Salai	8.453	5.556	- 34
9. Garari	2.697	6.626	+ 146
10. Other	18.079	22.467	+ 24
Total	63.150	64.375	+ 2

Source : Resource Survey (Working Plan of Rajnandgaon, 1999).

If we analyze the change of cop in teak forest with respect to girth classes, then one finds that there is a decrease of volume in the girth classes below 60 cmt. Likewise, there is an increase of volume in girth classes above 61 cmt (Table 3.21). In case of mixed forests, there is a decrease in volume in girth classes below 60 cmt. An increase in volume is witnessed in girth classes above 91 cmt (Table 3.22). These changes are in consonance with the management objectives.

Table 3.21 : Teak Forest Change in Trees and Volume : Girth Classwise

S. No.	Girth Class (cmt)	*No. of Trees per Hectare and Volume per Hectare*			
		1980 – Resource Survey		*1997 – Resource Survey*	
		Trees	*Volume*	*Trees*	*Volume*
1.	30 – 45	127.914	6.824	96.870	5.806
2.	45 – 60	40.285	6.555	42.292	4.643
3.	61 – 90	44.786	17.034	66.458	20.048
4.	91 – 120	17.325	14.675	22.083	15.085
5.	121 – 150	4.631	7.237	7.708	9.611
6.	Above 150	1.846	5.91	4.584	7.602
	Total	236.787	58.235	239.995	62.796

Source : Resource Survey (Working Plan of Rajnandgaon, 1999).

Table 3.22 : Mixed Forest Change in Trees and Volume: Girth Class wise

S. No.	Girth Class (cmt)	*No. of Trees per Hectare and Volume per Hectare*			
		1980 – Resource Survey		*1997 – Resource Survey*	
		Trees	*Volume*	*Trees*	*Volume*
1.	30 – 45	142.862	7.609	103.11	6.446
2.	45 – 60	47.194	7.618	55.672	6.267
3.	61 – 90	48.663	18.006	61.489	18.084
4.	91 – 120	15.909	13.275	22.607	15.369
5.	121 – 150	5.078	7.442	8.952	10.978
6.	Above 150	2.468	9.200	4.538	7.232
	Total	262.144	63.150	256.373	64.796

Source : Resource Survey (Working Plan of Rajnandgaon:1999).

However, two things stand conspicuously. First is the increase in volume of saja species and second is increase in saja is seen of the girth range of 90 to 150 cm only.

Change in Growing Stock

During the last 17 to 20 years, there is slight change in quality of growing stock. The first inventory was conducted in the year 1981. On the basis of that inventory, the growing stock was measured and this constituted the base line data for comparison. During the revision of working plan process, the growing stock was again measured. Both the data can now be compared and analysed.

The growing stock of teak forests shows almost no change. In the year 1981 the growing stock was 10.983 lakh cubic metre. In the year 1997, it decreased to 10.570 lakh cubic metre. This change is very marginal. What is striking is the fact that mean volume of trees per hectares in teak forests increased by about 10 per cent, but, due to shrinkage of teak forest area, the growing stock declined marginally (Table 3.23).

The growing stock of mixed forests dropped sharply by about five per cent. What is striking is the fact the mean volume of trees per hectares increased marginally. But, because the extent of mixed forests declined sharply, especially the dense mixed forests, increase in mean volume of trees per hectares did not reflect in the growing stock (Table 3.23).

If we analyse the change in growing stock of the study area in its totality, then it is found that there is a decrease in growing stock by about 5.6 per cent in a span of 16 years. In the year 1981, the total growing stock was 138.377 lakh cmt and in the year 1997, it came down to 130.601 lakh cmt, thereby a fall of about 6 per cent (Table 3.24).

Table 3.23 : Growing Stock of Forests :Year : 1980 and 1997

Growing Stock in year 1980

S. No.	*Forest Type*	*Forest Area*	*Volume per hectares (cmt)*	*Growing Stock in lakh (cmt)*
1.	Teak Forest	18860.24	58.235	10.983
2.	Mixed Forest	201732.96	63.150	127.394
	Total	225916.90		138.377

Growing Stock in the : Year 1997

S. No.	*Forest Type*	*Forest Area*	*Volume per hectares (cmt)*	*Growing Stock in lakh (cmt)*
1.	Teak Forest	17397.243	62.796	10.570
2.	Mixed Forest	186068.384	64.375	120.031
	Total	203465.627		130.601

Source : Resource Survey (Working Plan of Rajnandgaon, 1999).

Growing Stock of Bamboo

Bamboo constitutes an important resource in the study area. The growing stock of bamboo forests reflects a harrowing trend. In the study area, the extent of bamboo forests is 61499.291 hectares. Out of it, the first class, second class and third class bamboo forests constitute about 35 per cent, 50 per cent and 15 per cent of the total bamboo forests respectively. The growing stock of bamboo constitutes about 856 lakh of bamboo culm equivalent to 178333 NT (Table 3.24).

Table 3.24 : Growing Stock of Bamboo Year 1997

S.No.	*Nature of Bamboo Culms*	*Number of Bamboo Culms (in lakh)*
1.	Ist Year (Karla)	138.85
2.	IInd Year (Mohila)	239.60
3.	IIIrd Year (Pakiya)	477.57
	Total	856.02

Note : The average length of the bamboo is 5 mt. and 2400 metres length of bamboo constitutes one Notional Tonne (NT).

Source : Resource Survey (Working Plan of Rajnandgaon, 1999).

During the last plan period there was a shrinkage of 3920 hectares of bamboo forests. Taking 1432 Bamboo (average) per hectares, the loss of growing stock comes to about 56 lakh bamboo culms. The annual depletion of growing stock is about 700 NT.

Change in Status of Regeneration

Regeneration of species holds the key to sustainability of ecosystem. Without sufficient natural regeneration, it is impossible to hold on the forests for a long time. Artificial regeneration of plantation is a tool to rehabilitate degraded forests. But there is a limit to artificial regeneration. In case of the forests of the study area, natural regeneration is not satisfactory (Table 3.25). Majority of the area shows strained pattern of natural regeneration. Majority area in different working circles shows distribution of regeneration as poor to very deficient (Table 3.26).

Table 3.25 : Status of Natural Regeneration in Different Proposed Working Circle

S.No.	*Working Circle*	*No. of Regeneration per Hectare*			
		Teak	*Mixed*	*Total*	*Remarks*
1.	Selection-cum-Improvement	68.88	1067.12	1136.01	Deficient
2.	Improvement	56.15	1023.30	1079.45	Deficient
3.	Rehabilitation of Degraded Forest	22.46	631.42	653.88	Very Deficient

Table 3.26 : Distribution of Natural Regeneration within Different Proposed Working Circle Areas

S.No.	*Proposed Working Circle*	*Number of seedlings per hectares*	*Grade*	*Area in %*
1.	Selection-cum-Improvement	2500>	Sufficient	4.60
		1501 to 2500	Poor	39.23
		1000 to 1500	Deficient	22.31
		<1000	Very Deficient	33.86
2.	Improvement	2500>	Sufficient	2.90
		1501 to 2500	Poor	23.84
		1000 to 1500	Deficient	27.33
		<1000	Very Deficient	45.93
3.	Rehabilitation of Degraded Forests (RDF)	2500>	Sufficient	0.00
		1501 to 2500	Poor	9.76
		1000 to 1500	Deficient	21.95
		<1000	Very Deficient	68.29

Regeneration of many tree species having important non-wood forest product is not encouraging. Take for example the case of achar or chironji. The stem distribution is not normal. This is true in other species also. All this highlights the elements of unsustainability in stand dynamics (Table 2.27).

Change in Productivity

Felling of trees is an important function of production forestry. It improves the light availability on the floor of the forests. It indirectly promotes not only seed germination but also stand development. Therefore, every working plan geared for production forestry emphasizes on felling of matured trees and improvement of crop. In the study area, the degree of felling in a unit area was analyzed by examining the felling data of the coupes in the last plan period (Tables 3.28, 3.29, 3.30 and 3.31).

Table: 3.27 Number of Important NWFP Trees per Hectare in the Planned area

Species	*11 – 20*	*21 – 30*	*31 – 45*	*46 – 60*	*61 – 90*	*91 – 120*	*121 – 150*	*151 – 180*	*Above 180*	*Percentage*
Tendu	1.49	3.85	2.61	1.53	1.86	0.95	0.55	0.22	0.04	3.26
Harra	0.60	1.37	1.18	0.78	1.08	0.42	0.33	0.04	0.09	1.46
Achar	1.57	5.49	5.39	2.41	1.53	0.20	0.02	0.00	0.00	4.13
Mohua	0.47	1.95	1.33	0.64	1.33	0.95	0.67	0.20	0.24	1.94
Baheda	0.04	0.15	0.29	0.20	0.13	0.13	0.04	0.07	0.00	0.26
Amla	0.67	3.12	3.88	1.49	1.66	0.31	0.06	0.00	0.00	2.78
Total	4.85	15.91	14.69	7.052	7.58	2.95	1.66	0.53	0.36	13.83

Source : Forest Resource Survey, 1997.

Table 3.28 : Felling Intensity : Species-wise Number of Trees per Hectare

S.No.	*Working circle*	*No. of trees*					
		Teak	*Saja/ Salia*	*Bija/Tinsa/ Sisam*	*Haldu/Mundi/ Tendu/Dhawda*	*Other*	*Total*
1.	SCI / Conversion to Uniform/ Plantation	2.124	4.862	2.289	4.666	48.306	62.247
2.	CWR	3.239	10.439	2.855	36.685	50.313	103.74

Table 3.29 : Timber and Bamboo: Production: period 90-97

S. No.	*Year*	*Timber (cmt)*	*Fuel wood (cmt)*	*Bamboo (NT)*	
				Commercial	*Industrial*
1.	1990-91	24356.798	49845	11133.216	25523.532
2.	1991-92	28588.902	65743	11346.553	21704.720
3.	1992-93	30278.890	53266	18387.987	20077.787
4.	1993-94	22140.449	41998	13483.328	19098.973
5.	1994-95	20239.918	34526	12189.735	22814.046
6.	1995-96	12675.224	23407	7312.236	15432.342
7.	1996-97	16350.674	29748	12472.873	21626.228

Note : Total Exploitable of bamboo forest was 41188 hect.

Table 3.30 : Timber and Bamboo Productivity Trend: Period 90-97

S.No.	*Year*	*Timber & Fuelwood production in cmt*	*Annual Timber Productivity per hectares in cmt*	*Bamboo Production in NT*	*Annual Bamboo Productivity per hectares in NT.*
1.	1990-91	74202	0.62	36657	0.89
2.	1991-92	94332	0.78	33052	0.81
3.	1992-93	83545	0.69	38467	0.94
4.	1993-94	64138	0.53	32582	0.80
5.	1994-95	54766	0.45	35004	0.85
6.	1995-96	36082	0.30	22744	0.56
7.	1996-97	46099	0.38	34099	0.83

Note : Sixty per cent of forest is taken as workable area and eighty per cent of bamboo area workable.

There is no perceptible change in productivity of tendu leaves, myrobalans, gums and other non-wood forest products. However the population of tendu trees in the study area witnessed a sharp decline. Though there is no direct relationship between production of Tendu leaves and the tendu tree population, but the phenomenon is a disturbing one. Tendu leaves are plucked from root sucker seedlings, which come normally in the fringe areas of the forests. Declining tree population is definitely a potent indicator of ecological change. It may affect the sustainability of tendu leave production and trade in the study area. Another point that emerged from the study that the reason for high production of tendu leave in the study area lies in fact that the forests of the division witnesses a high degree of fragmentation. This has resulted into an enormous area of forest edge and that too in degraded forests.

The ecological role of tendu trees in mixed forests is not fully known. But the tree has a definite role in the forest ecosystem. Declining population of tendu trees is a cause of concern from the point of ecosystem management. Other species which are facing a crisis of regeneration are harra, achar, mohua, baheda and amla. Though the populations of these species except tendu have remained almost constant, but witnessed very scanty regeneration. The primary reason for lack of regeneration and declining population is uncontrolled fire, grazing and destructive harvesting practices by local people.

Table 3.31 : NWFP – Production Profile

S.No.	*Year*	*Tendupatta (Standard Bag)*	*Harra (Qt)*	*Gond (Qt)*
1.	1990-91	227015	3307.84	394.00
2.	1991-92	177129	2764.35	84.00
3.	1992-93	181644	3288.76	863.13
4.	1993-94	164110	1190.91	120.00
5.	1994-95	125955	416.74	176.00
6.	1995-96	184496	3112.65	227.91

Bamboo is another species which is showing declining trend in productivity. The bamboo area is shrinking. The quality and integrity of bamboo forests is declining. The productivity along with the quality of the product is also falling.

Change in Plantation Area

Artificial regeneration or plantation plays an important role in forest development. At current estimate, about 11 per cent of the forests of the study area are degraded. These areas need to be regenerated artificially. Beside this, plantations have been raised since the year 1961 and the process still continues today. A per the records, there are about 27464.47 hectares of plantation in the study area. A large chunk of plantation has been raised in ranges like Dongargarh and Gandai in the period of 1991-95. An account of plantation in the study area in different years as well as different plan is presented in Table 3.2, Table 3.3 and Table 3.4.

Table 3.32 : Extent of Plantation in the Study Area : Period 1961-2002

S.No.	*Range*	*Plantation in hectares*
1.	Chuikhadan	1545.00
2.	Chowki	6613.99
3.	Dongargarh	10553.80
4.	Manpur	1868.12
5.	Panabaras	400.100
6.	Rajnandgaon	2305.98
7.	Khairagarh	1175.00
8.	Gandai	30002.40
	Total	27464.470

Source : Forest Department's record submitted to Honourable Supreme Court : Year 2003.

Table 3.33 : Extent of Plantation in Different Period : 1961-2002

S.No.	*Period*	*Plantation in hectares*
1.	1961-65	344.05
2.	1966-70	655.00
3.	1971-75	937.05
4.	1976-80	1818.00
5.	1981-85	1931.13
6.	1986-90	2209.00
7.	1991-95	15563.50
8.	1996-2000	1976.93
9.	2001-2002	1347.00
	Total	27464.47

Source : Forest Department's record submitted to Honourable Supreme Court : Year 2003.

Table 3.34 : Plantation in Different Working Plans : Period 1981–1996

Plantation in different Working Circles (under Goyal's Plan upto 1995)					
S.N.	*Working Circle*	*Teak*	*Bamboo*	*Mixed*	*Total*
1.	Selection-cum-Improvement	356.72	429.00	1657.05	2442.77
2.	Soil and Water Conservation	693.40	844.00	5584.40	7121.80
3.	CWR	0.50	130.00	3362.00	3492.50
4.	Conservation to Uniform	308.00	70.00	158.20	536.20
5.	Plantation	1648.88	328.08	375.00	2351.96
	Total	3007.50	1801.08	11136.65	15945.23

(contd....)

(contd....)

Plantation in Different Working Circles (under Sahu's Plan upto 1996)

S. No.	*Working Circle*	*Teak*	*Bamboo*	*Mixed*	*Total*
1.	Improvement Felling	-	148.00	385.00	533.00
2.	Soil & Water Conservation	320.00	200.00	270.00	790.00
3.	Teak Plantation	74.00	-	230.00	304.00
4.	Energy Plantation	250.00	200.50	1306.00	1756.50
	Total	644.00	548.50	2191.00	3383.50
Grand Total Goyal & Sahu WP		3651.50	2349.58	13327.65	19328.73

Note : Other than the Forest Department, Forest Development Corporation (FDC) also raised teak and bamboo plantation in an area of 28154 hectares. These plantations are of best qualities.

A large area was brought under plantation during the period 1991-95. Out of three main categories of plantation, mixed plantation secured the prime importance. But, on assessment, mixed plantations were found least successful.

PART - II

Sustainable Forest Management (SFM) : Strategy and Options

4

Strategy for Sustainable Forest Management

Success in management depends in policies, legislation and strategies. A discussion on policies and legislation is out of the scope of this book. This chapter intends to discuss options available for strategic forest management on a board spectrum. Strategy plays an important role in forest management. It translates the policies and projects into action on the ground. It redeems the objectives of management. A strategy for sustainable forest management consists of broad direction for forest conservation, development, sustainable use, equity based benefit sharing, etc. The purpose of evolving strategies is not only to determine and communicate a picture of the kind of working that an organization visualizes, but also to develop practical framework for acting at ground level.

This chapter, while dealing with sustainable strategy option delves upon subjects such as three different models of forestry, principles of forest management at landscape level, policy support for management, strategy formation for forest modelling and activities under it.

Different Models of Forestry

Forestry in the broadest sense involves the science, art and business of managing forest for human benefit. Forestry began at different times and different places throughout the world as society's demand for wood outgrew the volumes obtainable by exploiting wild forests. From the perspective of use or non-use, forestry can be categorized into three forms. The first and earliest form of forestry could be characterized as custodian (focusing on protecting the forests from over exploitation and fire). This is also called preservation approach.

The second and the next form of forestry was characterized by sustained yield timber production (focusing or assuring a continuous supply of timber). Over the years, there is a gradual change in perspectives of this model too. There have been explicit efforts to manage forests for a broad array of resources. This led to evolution of multiple use forestry. This form of forestry is also called as utilization approach. This is the third form of forestry. Over the years, there is another thinking that we have entered into an era of ecological forestry or ecosystem management. In this form of forestry maintenance of ecological integrity remains para-mount. This is called as ecological approach. This is the fourth form of forestry.

Custodian forestry lost its relevance in the present time. It was supplanted by sustained yield timber production forestry. Key concepts of sustained yield forestry are rotation or exploitable girth, felling cycle, thinning regime and regulated age structure of the growing stock. Multiple use forestry received acceptance because of recognition of the fact that benefits from forests goes beyond timber and should, therefore, be managed for optimum benefits, tangible as well as intangible. With the passage of time forest management became synonymous with sustained yield-cum-multiple use forestry. This hybrid model aims at timber as well as multiple uses of forests. It followed commodity approach. It did not focus on managing the forest completely on timber or other goods and services. But, this lacked ecosystem perspectives. Many thought that the traditional model of forestry should be revamped to accommodate ecological concerns. This is how the ecology based approaches have started to have been adopted. Gradually in the modern times ecological approaches are intensively followed. This new approach is an incremental evolution of multiple use doctrine. This model is called as 'ecological forestry', 'forest ecosystem management', and 'new forestry'. What distinguishes ecological forestry, from other traditional models is the emphasis placed on natural patterns and process: understanding them, working in harmony with them, and maintaining their integrity, even when it becomes financially difficult or inconvenient to do so.

The new paradigm of ecology has changed the concept of nature in the ecology. A new way of thinking replaced the old concept of 'balance of nature' with new concept of 'ecosystem dynamics' or 'flux of nature'. According to (Pickett and Ostefelt, 1995) the classical paradigm leads to many false assumptions that undermine many, but not all of the principles that management system have based on. These are: (a) systems are closed to out side influences and therefore, management can ignore changes outside management boundaries; (b) system are self regulating, and will remain relatively stable in the face of environmental change, and'therefore, management relies on benign neglect; (c) system passes through a single end point at which they are at equilibrium and therefore a management strategy of benign neglect will allow systems to maintain stability or returns to the same compositional and functional state they were at in the past; (d) succession always proceeds through the same pathways, so temporal changes are predictable and will take care of themselves; (e) disturbance is something that is outside the system so management can ignore it or try to stop it if it occurs; and (f) humans are not components of ecosystems. The new paradigm of ecology 'flux of nature' or 'ecosystem dynamics' is based on the assumptions that: (a) ecosystems and landscapes are dynamic; (b) disturbance is a critical components of the system; (c) ecosystems are controlled by biotic and physical process that occur at different spatial scales and levels of the biological hierarchy; (d) succession does not necessarily follow the same path and end at the some equilibrium point; (e) spatial pattern is important to biological diversity; (f) pattern and process interactions are organisms specific; and (g) human activities of the recent and distant past have had strong influences on many ecosystems that we may perceive as 'natural' today (Pickett and Ostefelt,1995)

The present approach to forest management is a *hybrid one*. It takes few elements from all approaches. The utilitarian approach of forestry does not apply fully to the current situation. Beside serving the present generation, we need our forest to be conserved for the future too. Likewise, the preservational approach is not always applicable. We need to serve the present society too. Therefore, the mid-course is the sustainable ecosystem approach, which seeks to serve both the present need as well as future need with out impinging the resource base, structurally and functionally. This is the ecological approach and it is basically an amalgamation of two aforesaid approaches.

Principles of Forest Management

There are few broad principles required to be observed while managing forests sustainably. They relate to sustainability, forest land use planning, ecosystem based management approaches soil and water conservation, non-destructive harvesting and sustainable utilization, biodiversity conservation, etc. They, if followed, aim at producing good and services from forests for the welfare of the society in such a manner that the forest resource base is kept intact for the present as well as future generations. We will be discussing the following principles one by one :

- Sustainability
- Sustainable forest management
- Ecosystem Approach
- Forest Land Use Planning
- Soil and Water Conservation
- Non-destructive Harvesting Practices
- Biodiversity Conservation

Sustainability

Sustainability is a principle or value which guides and directs our thinking/decisions today. It is based on the premise that the needs of the present generation should be met without compromising the ability of the future generations to meet their own needs. It encompasses all aspects of productivity and equity in an 'inter-temporal' context. There are three sub-components of sustainability. They are stability, resilience and non-declining natural stock.

Stability symbolizes the elements of 'constancy' of capital as well as productivity of the biological production system. In other words, it is the ability of the system to maintain equilibrium in response to normal fluctuations in the environment. Coefficient of variation is the measure of stability. Stability indicates the state of dynamics in an ecosystem. Likewise resilience is the capacity of the biological system to maintain its structures and patterns of behaviour in the face of 'external disturbances' which, with passage of time, emerge as stress and shocks. Thus resilience is the ability of a forest ecosystem to withstand such prospective disturbing forces. Forces of resilience act against forces that try to disturb the system. In such cases, the eco-systems develop their own defense system. Such defence mechanism differs from one ecosystem to another. Non-declining natural capital stocks are those natural resources whose decline affects the productive capacity of the system. Soil, vegetation, water and biodiversity are the natural capital stock, which are required for ensuring productivity and sustainability of the forests.

Sustainability of forests is one among the basic principles forest management. However, there is much confusion over the precise meaning of "sustainability". It is always preferable to think of sustainability in the context of human use and the aspirations of people. In this context, FAO defines sustainable development as "the management and conservation of the natural resource base and the orientation of technological and institutional change so as to enable the attainment and continued satisfaction of human needs for present and future generations" (FAO, 1991). Similarly, the Brundtland Commission (WCED, 1987) says "sustainable development is development that meet the needs of the present without compromising the ability of future generations to meet their own needs".

In the forestry sector, this concept is particularly valid. As stated by FAO Forest Resources Division Director J.P. Lanly in a recent paper presented to the International Tropical Timber Organization: "It must be accepted from the start that utilization of a given forest ecosystem implies some change in its structure and composition, and that sustainability cannot mean the identical reproduction of the ecosystem in its original state. The same applies to the conservation of the genetic diversity of a given forest...." Lanly goes on to affirm that forestry activities must be designed and implemented so as to maintain the level of conservation contemplated in the objectives of management.

Sustainable Forest Management (SFM)

Sustainable forest management is the forestry component of sustainable development. It is about operationalization of the concept of sustainable development in the realm of forestry. In other words, it is managing forest development in such a way that it is economically viable, environmentally benign and socially beneficial and at the same time, it balances present and future needs. More precisely speaking, sustainable forest management is the process of managing forests to achieve one or more clearly specified objectives of management with regard to the production of a continuous flow of desired forest products and services, without undue reduction of its inherent values and future productivity and without undue undesirable effects on the physical and social environment (ITTO, 1992).

Defining and redefining sustainable forest management has evolved into a process. With the change in time and societal needs, more and more aspects relating to socio-economic and ecological imperatives have been added to precepts of sustainable development. This apart, the concept of sustainability and sustainable development is also evolving in the context of prevailing social and economic condition of the society. All such developmental concepts based on sustainability are integrated into forest management framework. Therefore, the concept of sustainable forest development put in a management framework is called as sustainable forest management.

A number of authors have recently attempted to define sustainable forest management in general terms. Some of the examples are as follows :

(i) Gale and Cordray (1991) raised the question 'What should forests sustain?' and indicated that the answer varied depending on:

- the maintenance of a dominant product or product mix;
- the sustenance of production capacity;
- the conservation of total forest ecosystems rather than specific components of such ecosystems;
- the maintenance of human systems that are forestry-dependent.

(ii) According to Poore (1993) the sustainable management of forests comprises three components :

- the condition of the forest must not deteriorate irreversibly,
- management of the forest should be environmentally and socially acceptable and economically viable,
- the process of management itself requires that a well defined policy must be in place with the resources and ability available to implement it.

(iii) The definition developed by Poore for the ITTO is as follows (ITTO, 1992) :

'Sustainable forest management is the process of managing permanent forest land to achieve one or more clearly specified objectives of management with regard to the production of a continuous flow of desired forest products and services without undue reduction of its inherent values and future productivity and without undue undesirable effects on the physical and social environment'.

(iv) Sustainability may also be viewed in terms of forest functions, which may include products, services and values. At the management unit level, management of a certain forest may be considered to be sustainable when it guarantees the continuity of all recognized principal functions of a particular forest without undue effects on other functions (DBB, 1994).

(v) Heuveldop (1994) defined sustainable forest management as follows :

> Sustainable forest management comprises all those direct and indirect measures of protection, tending and utilization, which ensure the permanent conservation of forests. Sustainable management maintains the natural plasticity and diversity of life forms, which enable forests to evolve and provide human benefits from their ecological, economic, social and cultural functions in perpetuity.

(vi) Upton and Bass (1995) defined sustainable forest management by the three dimensions of sustainability; these dimensions are : (1) environmental sustainability, (2) social sustainability, and (3) economic sustainability. These three dimensions are explained below:

- *Environmental Sustainability :* This entails an ecosystem being able to support healthy organisms, whilst maintaining its productivity, adaptability and capability for renewal; it requires that forest management respects and builds on natural processes.

- *Social Sustainability* : This reflects the relationship between development and social norms. An activity is socially sustainable if it conforms with social norms, or does not stretch them beyond a community's tolerance for change.

- *Economic Sustainability* : This requires that benefits to the group(s) in question exceed the costs incurred, and that some form of equivalent capital is handed down from one generation to the next.

(vii) United Nations Commission on Sustainable Development (UNCSD) has defined Sustainable Forest Management as : "management to meet the social, economical, ecological, cultural and spiritual needs of present and future generations. In other words, Sustainable Forest Management (SFM) aims at ensuring that the values derived from the forest meet present-day needs while at the same time ensuring their continued availability and contribution to long term development needs. The values can be forest products like wood, fruits, etc. and services like soil and water conservation and life supporting services. Forest management involves planning the production of wood for commercial purposes as well as meeting local needs for fuelwood, poles, food, fodder and other purposes. It includes the protection or setting aside of areas to be managed as plant or wildlife reserves, or for recreational or environmental purposes." (FAO, 1993). "Sustainable management is considered under the headings of wood conservation, wild-life management, the conservation of biological diversity, forest protection and climate change; policy, institutional and legal aspects, socio-economic aspects and research needs." (FAO, 1991)

The term 'Sustainable Forest Management' (SFM) has been in vogue since Rio Earth Summit in 1992. It is a widely used phrase in political, scientific, environmental, economic and social context. Many perspectives on SFM exist today. Some of the views about broad connotations of SFM are as follows :

- Producing continuous flow of desired forest products with out loss of inherent productivity.
- Maintaining and enhancing the ecosystem functions, without entailing production or extraction.
- Prudent management.

Ecosystem Approach

Ecosystem approach in forestry is a recent phenomenon. In the past, certain approaches of ecosystem management were followed while managing forest. But at present, the approach of landscape management has been widely accepted. In this context, it is pertinent to discuss what we mean by ecosystem approach.

The convention on Biological Diversity describes the ecosystem approach as follows: "The ecosystem approach is a strategy for the integrated management of land, water and living resources that promote conservation and sustainable use in an equitable way. An ecosystem approach is based on application of appropriate scientific methodologies focused on levels of biological organization, which encompasses the essential structures, processes, functions and interactions among organisms and their environment. It recognizes that humans, with their cultural diversity, are an integral part of many ecosystems."

Ecosystem approach heralds a new beginning in forest management. It marks a shift from production output from the forest under the concept of 'sustained yield management' to 'ecological processes' inherent in forest ecosystem. The principles and operational guidelines for the ecosystem approach better known as " the Malawi Principles" are presented in Box 4.1. Sustainable forest management from the point of ecosystem is an evolving option in the forestry front. Approaches to forest management such, as "ecosystem and landscape" management are becoming more widely accepted in many countries. These approaches recognize the dynamism of ecological and social systems, the necessity of adaptive management and the importance of collaborative decision-making.

Ecosystem management approaches are basically based on critical elements like recognition of ecological and social systems, ecologically derived boundaries, consideration of different time scale, adaptive management to deal with change, uncertainties and collaborative decision making. Some people views the maintenance or restoration of ecosystem integrity or health as the overall goal of management, where as other considers human needs to be equally or more important. Many countries are increasingly using an ecosystem approach for planning and managing natural resources. In addition, parties to the convention on Biological Diversity agreed in the fifth conference on May, 2000 to apply the principles of the ecosystem approach in managing their natural resources.

The moot point in ecosystem management is how to maintain the structural as well as functional integrity of the managed forests, *i.e.*

outside Protected Areas (PAs). A natural forest ecosystem always tries to conserve structural and functional integrity. Human intervention in the form of silvicultural regimes symbolizes a compromise between ecological integrity and societal demand for forest products and services. It is always difficult to balance these two conflicting demands in forest management practices. With the fast changes in political, social and economic milieu of the society, the balancing act is becoming more difficult. Now it is clear that this approach in forest management can satisfy the society's needs without impairing the potentiality of the forest to provide benefits in the future. Therefore, it is necessary to adopt this approach to achieve management objectives.

Forest Land Use Planning

Forest has scope for multiple uses. Based on site quality index, topography, land suitability in terms of productive potential of biomass per hectare, forest land can be categorized into different circles or broad zones. The management objective of a zone should be clear and must not conflict with other zones. However, each zone can have two sets of objectives: first, primary objective and second, secondary objective. For an example, a zone can have timber production as primary objective and watershed protection as secondary objective. The parameters for allocating forest compartment into different zones cannot be entirely economic or exclusively biological. Decision should be guided by the factors of comparative economic, social and environment advantages. In practice each working circle or zone in a given working plan unit should be permitted to have certain uses compatible with its character and environmental requirement.

The aforesaid principle of zonation/division holds good for the study area. This aims at ensuring that each zone or parcel of land should be used for the purpose for which it is best suited and that any changes in use be made with the least risk of reducing the land's potential to human welfare. Where there are conflicts of views about what the best use might be, as must inevitably happen from time to time, then it should be decided by taking a holistic view of resource distribution and availability as well as the social, economic and environmental requirement of the people. Any temptation to make rapid capital out of the exploitation of resources at the expense of future benefits should be resisted.

Box 4.1 : Principles and operational guidelines for the ecosystem approach

The Malawi Principles :

Principle 1 : The objectives of management of land, water and living resources are a matter of societal choice.

Principle 2 : Management should be decentralized to the lowest appropriate level.

Principle 3 : Ecosystem managers should consider the effects (actual or potential) of their activities on adjacent and other ecosystems.

Principle 4 : Recognizing potential gains from management, there is a need to understand and manage the ecosystem in an economic context. Any such ecosystem management should:

(a) Reduce those market distortions that adversely affect biological diversity;

(b) Align incentives to promote biological diversity conservation and sustainable use;

(c) Internalize costs and benefits in the given ecosystem to the extent feasible.

Principle 5 : A key feature of the ecosystem approach includes conservation of ecosystem structure and functioning.

Principle 6 : Ecosystem must be managed within the limits of their functioning.

Principle 7 : The ecosystem approach should be undertaken at the appropriate temporal scales.

Principle 8 : Recognizing the varying temporal scales and lag-effects that characterize ecosystem processes, objectives for ecosystem management should be set for the long term.

Principle 9 : Management must recognize that change is inevitable.

Principle 10 : The ecosystem approach should seek the appropriate balance between conservation and use biological diversity.

Principle 11 : The ecosystem approach should consider all forms of relevant information, including scientific and indigenous and local knowledge, innovations and practices.

Principle 12 : The ecosystem approach should involve all relevant sectors of society and scientific disciplines.

Operational Guidelines

1. Focus on the functions of biodiversity in ecosystems.
2. Promote the fair and equitable sharing of the benefits derived from the functions of biological diversity in ecosystems.
3. Use adaptive management practices.
4. Carry out management actions at the scale appropriate for the issue being addressed, with decentralization to the lowest level, as appropriate.
5. Ensure inter-sectoral co-operation

Great stress is required to be laid on the value of the undisturbed natural forests as resources. It should be considered an essential part of land use planning to make adequate provision for the conservation of sufficient samples of each kind of forest or other natural ecosystem for the preservation of genetic resources. In addition to this, there are other advantages in maintaining further areas under a cover of natural or real natural vegetation. These provide the best and cheapest form of protection of the under laying soil and of the quality and regularity of flow of water form catchment areas.

Soil and Water Conservation

Soil and water conservation is one of the most important services forest provides to earth. The study area has many tributaries that originate from the forests. It has tributaries joining Wainganga which finally joins into the river Godavari. What is again more important is the fact that the forest-clad mountains are all situated in the western part of the district and the agricultural land lies in the eastern part of the district. Most of the mountains slope towards east. Therefore, the water in the river or nala flows towards eastern part of the district, which encompasses the agricultural field. Hence, sustenance of agricultural growth and productivity depends largely on the continuance of water flows in the rivers, streams and nalas. Considered from this perspective, sustaining the forest in the mountains is imperative. This is possible if we protect the soil resource of these mountains and maintain the site quality of the forests. Otherwise, the economy of the study area, which is critically dependent on agriculture and forestry, may collapse. Therefore, at least so far the study area is concerned, forest plays an important role in sustaining human well-being of the district.

The methods of conservation should be so developed that it takes into account the need to protect the stability of the system in area where disturbance may cause accelerated soil erosion, unfavourable change in stream flow or loss of structure and lead local soil disturbance. The degree of soil erosion, and of reduction in water infiltration depends not only on the slope, soil structure, soil depth and rainfall intensity but also on the size of the area affected and the time taken to restore a vegetative cover. Whenever soil and water resources are vulnerable to changes in forest structure, the retention of continuous cover over the land should be the approach.

Managing forests for conserving soil and water is a major concern of the planner. It is carried out for many purposes or for combining several complementary or conflicting objectives. Scenarios range from cases where the only requirement is to establish norms for minimizing damage to soil and water resource by observing a few basic management rules – including road construction, land preparation and forest harvesting – to cases, such as critical forest watershed, where the need for undisturbed forest cover may be so overriding as to exclude all activities that would affect or change the forest vegetative cover. This is achieved by adopting two pronged strategy : (1) by taking steps for soil and water conservation in erosion prone areas, and (2) by excluding certain prone areas from timber harvesting or minor forest produce collection.

Sustainable forest management for soil and water conservation cannot be separated from the social and economic conditions prevailing in the society. The first guiding principle should, therefore, be to consider forests in the national and local socio-economic context. Second, it is worthwhile to consider a range of scenarios such as : (1) natural forests should be protected in its natural state for water, wildlife, and genetic resource conservation; (2) natural forests should be subjected to wood and non-wood forest products harvesting ; (3)forest area should be taken up for raising plantations. Each of these needs specific action. The positive or negative effect of logging or harvesting in managed forests on soil and water conservation will depend on the location of roads, the quality of their construction, quality of their maintenance, drainage; the type of felling and hauling operations and degree of mechanization used; the silvicultural interventions, felling or harvesting intensity and cycle. Therefore, the approach in management should be to consider a range of scenario from the point of soil and water conservation.

Non-destructive Harvesting Practices

Harvesting practices play an important role in sustainable forest management. Here harvesting includes "harvesting" as well as "logging". As used here, logging refers simply to the process of felling and extracting wood from forests whereas harvesting includes pre-harvest planning, technical supervision and post-harvest assessments that reflect concern about non-timber resource values and the future state of the forest. "Harvesting Technology" is a collective term that refers to the use of scientific and engineering principles in combination with education and training to improve the application of labour, equipment and operating methods in the harvesting of industrial timber. The key to promoting sustainability of forests during wood harvesting operations is to utilize the best knowledge available regarding five critical elements: harvest planning; forest roads; felling; skidding and stacking; and post-harvest assessments.

It is essential that efforts to sustain forests must proceed simultaneously along two fronts: the socio-political front which endeavours to develop economic, social and political conditions that will permit forests to be sustained; and the technological front which seeks to ensure the use of harvesting practices that are environmentally sound and economically acceptable and that simultaneously promote the sustainability of both the timber and non-timber values of the forest.

To contribute to sustainable development, then, a critical aspect of the utilization of forests is that the activities associated with the utilization must not irreversibly compromise the potential of the forest to regenerate and continue to provide the wood and non-wood forest products, environmental services, social benefits including maintenance of biodiversity essential for the well-being of both current and future generations. In fact, the goal should be to maximize the aggregate potential of the forest to provide these goods and services over the long term. This implies that harvesting operations should be carried out in such a way as to leave the forest in a condition that favours a rapid recovery to its pre-harvest state or to some other state that is silviculturally, ecologically and sociologically desirable.

Biodiversity Conservation

Forest resources, like all resources produced by living systems, are renewable provided their "biological capital" remains intact. This capital includes goods and services provided by the growing stock, for example timber or watershed protection, and also overall variety and variability. Although there is a danger of reducing the complexity of the human-forest codependency to a purely economic or production perspective, the consequences of deforestation can be disastrous.

The biological capital, including the overall variety and variability of forest ecosystem is biodiversity. This diversity should be conserved at any cost. Biodiversity encompasses all three levels: ecosystem level, gene level, and species levels. Attributes of variety and variability contribute to the forces of resilience, evolutionary potential and dynamic equilibrium found in the ecosystem. Consequently, all the forces strengthen the foundation of sustainability of forest ecosystem. The heterogeneous forest landscape mosaic buffers against the effects of natural threats such as inclement weather and pest outbreaks. Species competition and redundancy within ecosystems enhance species fitness and help ensure the resilience and genetic survival of a species in the face of environmental stresses of human and natural origin.

Although it is not certain how much variety and variability is necessary to provide specific goods and services, methods do exist to estimate requirements for maintaining the viability of valued plant and animal populations as well as the integrity of the underlying natural ecosystem. Therefore, while undertaking forest planning it is necessary to adopt biodiversity conservation measures.

Stress on forest ecosystem is increasing day by day. There are many sources of stress (See Box 4.2). But, the root cause of such stress lies in the society. The impact of such stress on the ecosystem is enormous. It degrades the forest ecosystem. It impinges on the diversity of forest ecosystem (See Box 4.3).

Box 4.2 : Sources of stress on a forest ecosystem

- Over-harvesting of renewable resources
- Pollution
- Physical restructuring
- Exotic species
- Natural variation

Box 4.3 : General process of forest ecosystem degradation

- Reduction in species diversity
- Depletion in the nutrient pool
- Change in structure from forest towards hardy shrubs and herbs
- Decrease in complexity of relationships between species
- Decrease in primary productivity

Source : Woodwell (1970).

Operationalising biodiversity conservation is a challenge. Again, within this framework, conserving genetic resource is more challenging. As human populations continue to increase, setting aside large areas purely to preserve biodiversity becomes more and more difficult. At the same time, however, the conservation of genetic resources is fundamental to the sustainable management of the forest ecosystem in which they occur. In situ conservation of biodiversity including forest genetic resources, and particularly those of trees which are dominant in determining the unique genetic potential of the system, should be reinforced in the management of production forests.

To a large extent, each forest area is unique in terms of its genetic resources and, consequently, in terms of its potential contribution to the achievement of forest production and genetic conservation objectives, as determined by prevailing ecological, social and economic needs. It is neither possible nor desirable to prescribe equal priority and intensity to *in-situ* conservation in all production forests. However, all management plans should include provisions for the protection of site conditions, seed trees, seedling regeneration and advance growth of desirable species.

Policy Support

Many tropical forestry experts agree that it will be impossible to sustain tropical forests over the long term without the establishment and enforcement of appropriate government policies. For instance, authorities such as the Brundtland Commission (WCED, 1987), FAO (1993) and, Poore (1989) suggest that policies must be developed to ensure comprehensive land-use planning; provide incentives for sustainable practices; encourage the increased use of agro-forestry and forest plantations; ensure the participation of local peoples in the benefits of timber harvesting; and protect the primary forest. After such policies have been developed, governments must also have the political will to enforce them; only then there will be a chance for tropical forests to be sustained.

Policy guidelines determine the management objectives, which finally decides strategy and activities. Since the study area is in the state of Chhattisgarh, the National as well as the State Forest Policy shall determine the management objectives. The State Forest Policy hierarchically supports the National Forest Policy, 1988 in letter and spirit. But it distinctly sets the priority for the state too. The salient features of the state forest policy are :

- Preeminence of ecological sustainability.
- Community driven forest management.
- Service to local people through eco-friendly Nistar policy.
- Forest as a mean to alleviate poverty.

- Restoration of degraded forest.
- Non-destructive harvesting.
- Sustainable management of existing growing stock.

Forest Planning at Landscape Level

Forest management requires decisions that are based on both scientific knowledge and social values. These decisions are influenced by many goals advocated by many parties. Forest planning is a mechanism for decision making that links social values with science. Forest management translates these values into activities that promote those values. Planning involves blending of ecological and social systems. Social side of planning is more complex, then the ecological side. Therefore, the integrated exercise of forest planning is an uphill task. One should remember that planning will not be successful if it flies in the face of the natural processes by which management actions are translated into ecological effects, nor will it be successful if it fails to produce the services and products that are desired by society.

The earth's natural resource base is finite. It has a definite carrying capacity. Use of natural resource beyond the threshold limit leads to impairment of the natural resource base. Therefore, there is a 'limit to growth' with respect to use of natural resources. Forest is an inalienable component of the earth's natural resource. With escalating human demand on a shrinking natural resource base, it is vital to ensure that planning for meeting human demand takes all aspects of the landscape together in totality. In other words, to deal with the range of essential needs, to integrate economic, environment, and social aspects of development, and to anticipate and deal with the feedback from one sector to another, we require a form of planning which takes all aspects of the landscape together in totality. Managing forests for sustainability must emerge from such considerations and fit in within a broader approach to planning for and managing the overall landscape. A process of planning must lead to identification of how the requirement of forests for conservation and of community for basic needs without entailing the future generation's potentiality to get at least same quantum of social welfare can be met. This method of planning must be conducted in conjunction with the context of other natural resources in the ecosystem. In other words, the planning should encompass all aspects of natural resource management in such a way that the execution of plan leads to increase in natural capital, including forest. Therefore, there is a need to embark on a pattern of holistic planning of the landscape, in which forest is a vital component.

Forest resources are scarce. They have definite carrying capacity. Using forest resources beyond a threshold limit leads to impairment of the resource base itself. Now the ticklish question is how to plan the use of forest within the threshold limit. If we want to find out a answer to this question, then we should begin with planning at landscape level (See Box 4.4). Forest planning at landscape level is not a new thing. It has been practiced in the past also. Landscape planning, basically, aims at meeting human needs taking all aspects of the landscape together in totality. In other words, it not only ensues meeting essential needs but also integrates economic, environmental and social aspects of development. Therefore, landscape planning is a process of planning by which we find out ways and means of meeting the needs of the people without jeopardizing the interests of future generation or entailing the future generation's potentiality to get at least same quantity of social welfare.

Box 4.4 : Requirements of Landscape Planning

Three elements of planning are essential for environmentally sound and socially acceptable forest management :

- That areas of forest to provide for all these uses should be set in wider, national or bio-regional, land-use context which provides adequately for all of them in suitable locations;
- That forests should be available to meet the various needs of society; and
- That each areas of forest are carefully managed for the purpose for which it is suited and intended.

The exact details will very greatly from country according to the area and nature of its forests, its population density, its state of development and many other conditions. This pattern should retain or even enhance the 'potential' of forests and give them the flexibility to respond to changing conditions whether social, economic or ecological. The best use of a forest cannot be considered in isolation. For example, forest plantations play an important part in producing timber and may thereby reduce pressures on natural forests; if established on degraded lands they may also increase biological diversity. In general, planning for agricultural expansion, for new settlements and for roads should take into account the possible effects of these developments on adjacent forests.

Source : World Bank, 1978.

Planning for forest management is a process of observation, appraisal and analysis with respect to resources, organizational strength, problems/constraints and compatible courses of actions. This process is learnt much better if the results of thinking and reasoning are written down in a simple but organized way. Apart from this, the fact that sustainable development of forestry, which needs a long term planning, calls for a systematic, logical scientific and analytical plan. Such plans are executed for a longer period of 10 to 20 years. Therefore, considered in the backdrop of those factors, it becomes imperative to have an elaborate written plan in the field of forestry.

Societies demand more and more from forests to satisfy increasing needs. It is, therefore, essential to establish priorities, set in order of importance to their functions and to harmonize them. The purpose of forest management planning is to select the targets to be attained and to programme in the form of host of integrated interventions (silvicultural and others) at stand level to attain them. It seeks to achieve a balance between productivity and conservation.

Management Strategy

In order to develop the forest management strategy, one needs to undertake the following exercises :

1. Analysis (what can be done).
2. Strategy Development (what's intended).
3. Programme Formulation (what must be done).

Analysis

Analysis constitutes an important step in formulation of management plan. It includes all aspects of problems, strengths, objectives, inputs, internal and external influencing factors, organisational issues. The study area has its strengths and weaknesses. Unlike these aspects an analysed, a programmatic strategy cannot be developed.

Each forest has its own uniqueness in terms of vegetation, topography, soil resources, and microclimate. It calls upon to play a special role in its economic environment. Therefore, the first step is to survey the state of the ground and assess previous management. There is a need to analyze the existing situations make a quantitative and qualitative inventory of the stands, gather opinion and identify the local needs that forest can satisfy.

Aspects such as forests resources, factors of locality, external influencing agencies, of the study area have been discussed thread bare in the previous chapter. An inventory of resources has also been prepared during the course of forest resource survey. An analysis of inventory has been presented in chapter two and three.

Strategy Development

Most of the time the same forest is intended to fulfil several functions. This multiple use can generally be achieved within single forest. However, priority must be set and uses harmonized. Some functions may be conflicting or difficult to reconcile. In such situation, the forest must be divided into separate management units or circle or zones. More generally the forester must try to administer balances between protection and conservation, production and biodiversity conservation and etc.

Again under each zone or working circle area, management must try at spatial as well as temporal scale. The guidelines regarding rotation age, felling cycle, and exploitable girth determine the temporal scale. Likewise, arrangement such as felling series, coupe specific tree level felling and, and stand level felling determines spatial scales. Again, each zone or working circle requires determination of temporal scales at two levels. First is at stand level and second is tree level. Felling cycle or rotation denotes the temporal scale at stand and tree levels. The exploitation girth indicates the temporal scale at tree level. Temporal scales at both these levels should be decided in the strategy. Both temporal and spatial operations should be properly framed, scheduled and harmonized to provide an integrated spatial and temporal impact on the forests.

Programme Development

After having developed the right strategy, it is necessary to formulate a detailed programme of action. Programme of action shall encompass schedule of operations with respect to felling, other tending operations, subsidiary silvicultural operations and monitoring and evaluation system. All these operations shall be executed in working circles, felling or treatment area, coupes (annual treatment area). In order to translate strategy into programme of action, two types of measures can be taken up. They are as follows :

1. *Silvicultural felling and subsidiary silvicultural operations :* Management outlines a group of coupes and prepares a schedule, which plans the felling in time (year by year) and in space (coupe by coupe).
2. *Miscellaneous Regulatory Operation*: Regulations can impose some public interest constraints on the forest manager with regard to aspects of miscellaneous nature such as setting up a set of guidelines about felling, reserving an area for game reserve, protecting a site for its natural heritage value, maintaining seed production area, maintenance of boundaries, etc. Regulatory mechanism is necessary to effectively execute the silvicultural strategy. This imposes certain regulation on the behaviour of organization and public at large. An account of silvicultural management and miscellaneous regulation has been presented in Table 4.1.

Table 4.1 : Aspects of Silvicultural Management

Sl.No.	*Silvicultural Management*	*Aspects*
1.	Silvicultural System	Rotation Age Exploitable Girth Felling Cycle Felling Series Felling Coupes
2.	Miscellaneous Regulation	Wildlife Regulations Participatory Forest Management Forest Extension Nurseries Maintenance of Boundaries Nistar and Paidawar Forest Protection Sample Plots Irregular Exploitation Forest Land Transfer

Components of the Strategy

Forest management has a broad ambit. It is not possible to deal with all the aspects. However, discussion warrants a succinct mention about different components of the strategy. Forest management strategy encompasses mainly two parts. Firstly, it is socio-economic strategy and secondly, it is silvicultural strategy. The socio-economic strategy is generally prepared for a region or a state. It contains many components. The driving force for developing a socio-economic strategy originates from the problems of the society at the socio-economic level. Unless these socio-economic problems are not solved in right time, they start impinging on the silvicultural domain of forests. A detailed discussion on socio-economic strategy is out of the scope of this study. However, a list of components of socio-economic strategy and their driving force is presented in Table 4.2.

The second type of strategy is silvicultural strategy. Under this strategy, the forests of study area have been divided into six working circles or zones. Each working circle will have a set of harmonious objectives and a single silvicultural system. A particular silvicultural regime shall be developed for that circle. The regime shall symbolize a host of silvicultural interventions necessary for sustainable management of forests. Rotation, felling cycle and felling series differ from one circle to another. Detailed discussion on the silvicultural strategy has been made in the next chapter.

Few components of forest strategy, which warrant priority, need to be explained. While executing the plan, due emphasis should be given to these aspects.

Management of Existing Forests

Preeminence of ecological sustainability is the guiding principle of management of forests in the state. The forest resources base should remain intact. The other dimensions of sustainability such as social and economic are linked to the very existence of resource base, *i.e.* natural capital. Considered in this backdrop, the management plan or working plan for the study area has been prepared.

Community Driven Forest Management

Managing forests of the study area on sustainable basis has been a challenge. It is possible if we allow participation of people in the full spectrum of forest management activities. Joint Forest Management forms the basis of forest management in a couple of village committees the study area. There is participation at levels of decision making by the said committee members. Adaptive management practices are long followed. They have the intrinsic capacity to reconcile the dichotomy of threat perception arising out of conservation-development orthodoxy by taking into account the human sensitivities like their socio-cultural norms, beliefs and system borne out of history, culture and traditions.

Table 4.2 : Sub-components of Socio-economic Strategy and Driving Force

S.No.	*Components of Socio-economic Strategy*	*Driving force*
1.	Combating poverty	Unemployment Low income
2.	Demographic dynamics and sustainability	Population growth rate
3.	Promoting education, public awareness and training	Primary education, Adult literacy rate, Low resource awareness
4.	Changing consumption pattern	Annual wood availability, and consumption, Annual timber consumption
5.	Protection of non-forest watershed and quality and supply of freshwater	Agricultural intensification, Annual withdrawal of surface water, Per capita consump-tion of domestic water in urban areas
6.	Promoting sustainable agriculture and rural development	Use of pesticides, Use of fertilizers, per cent of arable land under irrigation, Energy use in agriculture
7.	Integrating Environment and Development in decision in decision-making	Lack of mindset, Lack of commitment to sustainable development.
8.	Information for decision making	Lack of will power, Lack of knowledge, Lack of proper data
9.	Combating deforestation	Increase in utilization of forest products, Wood harvesting intensity, Uncontrolled grazing, Increased fire incidences
10.	Conservation of Biological Diversity	Population explosion, Increased consumption, Rising standard of living

Rights and Concessions

Rights and concessions are some of the important social instruments by which community can be associated in forest management process.

Therefore, forest of the state recognize traditional rights and concessions of entry into forests and use of the produce thereof, by the people living in and around the forests. The provision of Nistar rights and concession remains a potent socio-economic instrument of forest management in the study area. The domestic requirements of fuel wood, fodder, minor forest produce and construction timber have been increasing over the years, leaving a negative impact on the forest resource base. Therefore, the plan has laid out two-pronged strategy to administrator rights and concessions at one end and promote correction and development of resources at the other.

Management of Ecotone Forests

The study area has at least two pockets of ecotone areas requiring special management practices. Such forests are not only ecologically sensitive but also provide basic needs that constitute the essential element of the livelihood security of the poor people. Such forests are found in the northern boarder, from where the sal forests begin and in south part where the miscellaneous forests meets the teak forests of Maharaṣhtra. These ecotone forests have been taken care of, and appropriate silvicultural treatment has been proposed.

Conservation of MFP including Medicinal Plant

MFP like Tendu Patta, Chironji, Kullu, Honey, Harra, Baheda, etc. are the essential elements of the means of livelihood of the tribal and poor people. NWFPs provide a range of goods for domestic use and as well as income generation by selling in the market. They act a source of emergency sustenance in times of hardship – when crop fails. They also form a primary source of income to the poor. Appropriate treatments have been suggested for in-situ and ex-situ conservation, domestication and non-destructive harvesting of NWFPs including medicinal plants with active support from local people.

Forest Protection

Forests, being an open access resource (OAR) are vulnerable to various kinds of negative practices such as theft, illegal grazing and encroachment. Protecting forests holds the key to ecological sustainability. The strategy focuses on protection mechanisms.

Controlling of Forest Land Diversion

The Forest Conservation Act, 1980 and the National as well as State Forest Policy vow for total regulation of forest land. Policy enunciates that diversion of forest land for any non-forest purpose should be subjected to a careful examination by specialists from the standpoint of ecological, environmental and social costs and benefits. Procedure for doing this has been suggested in the working plan.

5

Silvicultural Management: Objectives, Options and Basis

Concern for the sustainable management of forest resources has never been greater than today. Across the globe, numerous initiatives are under way to promote management plans aimed at maximizing the contribution of forests to the achievement of development and socio-economic goals without compromising the present and future condition of the forest resource base. Silviculture, as element of forestry has the daunting responsibility of providing the biological and technical options to achieve management objectives. Therefore, the aspect of silvicultural management plays vital role in developing a management plan for sustainable forest development.

This chapter delves upon the concept of silviculture, factors of silviculture management, and selection of silvicultural systems. It discusses various management options in the context of the forest of the study area and the process involving the constitution of the working circles. Management objectives and characteristics of each working circle have been dealt threadbare in the chapter. In the end, regulation and management of important miscellaneous subjects like rights and concessions, management of ecotone area etc. have been discussed.

Silviculture

Silviculture is basically concerned with the technology of forest crop production. It is an applied science, which rests upon fundamental elements of natural and social science. It has been variously defined as: the art of producing and tending forests; the application of knowledge of silvics in the treatment of a forest; the theory and practice of controlling forest establishment, composition, structure, and growth.

(Spurr,1979). Silvicultural practice consists of the various treatments that may be applied to forest stands to maintain and enhance their utility for any purpose. The practice of silviculture is concerned with the social as well as the biological aspects of forestry. The implicit objective of forestry is to make forests useful to society. Since forest management aims at maximizing the social benefits in a welfare society, silvicultural management aims at managing silviculture in such a manner that the broad objective of management is achieved.

The broad objective of management is to grow and maintain forests for goods and services. Silviculture of timber production is usually the most intricate because both the kind and quality of trees are of greater concern than with other forest uses. Silviculture aimed at managing forests for achieving multiple objectives such as timber production, NWFP production, soil and water conservation, wildlife management, and recreation management is usually more complicated. However, what is important is the fact that objectives of silvicultural management should be clearly defined and silviculture should be directed to their attainment. The implicit objective of all silvicultural management is for the good of the forests as an entity, not as an end in itself but as a means of ensuring that it will be a permanently productive source of goods and benefits.

Objective and Factors of Silvicultural Management

Human purpose is introduced by preferences for certain kinds of trees or stand structure that have desirable characteristics. Natural processes are deliberately guided to produce forests that are more useful than those of nature, and to do so in less time. The rate at which value is produced is the most important consideration. The quality of value can be another consideration. All these considerations can be taken care of in silvicultural management. Some of the basic objectives of silvicultural management are as follows :

1. *Control of stand structure*: Silviculture aims at design of stands with outward and internal shape that will serve the intended purposes. The shapes and sizes of stands can be altered on the basis of purposes. The internal structure of a stand is determined by considerations such as variation in species and age classes, the different layers or stories of vegetation and distribution of diameter classes.

2. *Control of composition*: One of the objectives of silviculture is to develop the composition of stands in such a way that it will be most suitable from economic, social and biological points of view. The basic means by which tree composition is controlled is through regulating the kind and degree of disturbance during the periods when new stands are being established. This is accomplished by the means of tending and other subsidiary silvicultural operations. Desirable species can be promoted by resorting to artificial regeneration also.
3. *Control of stand density*: Control of stand density is a function of silviculture. Very high stand density causes the production to be distributed over so many individual trees so that none can grow at an optimum rate and many may decline in vigour. If stand density is too low, the trees may be too branchy or otherwise malformed and the unoccupied spaces are likely to be filled with unwanted vegetation. Unless stand density is carefully controlled at the time a stand is established or during its development, it is almost certain to deviate from optimum density level at some stages of its life.
4. *Restocking of degraded forest areas*: Degradation of forest on account of anthropogenic pressures is a continuing process. Therefore, re-stocking is imperative. This can be done either by natural re-generation or artificial re-generation.
5. *Protection of forests*. Forests are prone to damaging agencies such as insects, fungi, fire and human. Proper control of damaging agencies can result in further increases in production. Forest protection involves modification of silvicultural techniques.
6. *Control of rotation length*: Trees allowed to grow beyond the optimum do not continue to increase in value at rates sufficient to provide acceptable returns. The risk of decay may increase the possibility that they will decline in value. Proper regulation of stand density can shorten rotations by making the final harvest trees grow to the desired sizes at earlier ages.
7. *Conservation of site quality:* Maintenance of the productive capacity of the forests is the most important function of silviculture. The site is the total combination of the factors, living and non-living, of a place that determines this productivity of life. The site factor that is subjected to long-lasting harm is the soil. This is the most nearly non-renewable

of the resources used. Soil damage impairs the capacity of the site to yield benefits of the forests – flora and fauna, biodiversity and water.

Forest is a dynamic entity. It needs to be managed in such a way that the renewable natural resources would be available to humanity in perpetuity. Therefore, the forests are managed according to sound principle of silviculture. Silviculture as already said, is the art and science of cultivating forest crop. A forest crop requires tending, harvesting and replacement by new crops. These operations should be carried out according to certain silvicultural procedures. When a set of silvicultural procedure is initiated along with a homogenous set of objectives of management then it is called silvicultural system. Thus, silvicultural system has been defined as a method of silvicultural procedure worked out in accordance with accepted sets of silvicultural principles, by which crops constituting forests are tended, harvested and replaced by new crops of distinctive forms. In other words, the silvicultural system is a package of inputs applied to the forest crop throughout its life with an objective of giving the forest a predefined character. The selection of silvicultural system for any type of forests necessitates consideration of an array of factors. These factors influence the choice of silvicultural systems. Some of the important factors are : (1) Condition of regeneration and growth in the forests, (2) Species composition, (3) Nature of terrain and soil, (4) Protection considerations, (5) Nature of forest produces, (6) Nature of ecological services required, (7) Economic considerations, (8) Existing rights and privileges, (9) Aesthetic considerations.

Why to choose the Silvicultural System

Silvicultural system symbolises programme for the treatment of a stand during a rotation. It includes two aspects: (1) method of reproduction referring to the treatments of stand and site during the period of regeneration or establishment; (2) tending operation referring to treatments at other times during the rotation. Various tending operations may be carried out during transition from the regeneration stage to maturity. These are done to improve the existing stand, to regulate its growth, and to provide for early financial return, without affecting any regeneration. Although the silvicultural system includes both the method of reproduction and tending operations, it is customarily given the name of the reproduction employed. This practice is followed because the reproduction method has a decisive effect on the structure and treatment of the stand throughout its life.

The decisions made in silvicultural practice are based fully as much on economic and social objectives as on the ecological factors that govern the forests. Recognition of objectives and limitations set by societies in any given case reduces the silvicultural alternatives that need to be considered. Even though application of silviculture can make positive contributions to the forestry, the strategy for solving problems associated with social sciences is not a part silviculture. Such matters are dealt with in the field of forest management, which is concerned with both the organization of collections of stands into integrated forests and the administrative aspects of forestry (Davis, 1966).

Silvicultural management is based on the basic assumption that the vegetation on any site tends to extend itself aggressively to occupy the available growing space. The limit on growing space is usually set by the availability of light, water, nutrients, etc. In a general way the available amount of growing space is set by the most limiting factors. Although an abundant supply of one factor can partially offset deficiency of another. If vegetation fills the growing space, the only way that the forest can be controlled is by removing certain trees and promoting some. In regeneration felling, this is done to provide room for the establishment of new trees. Tending operation is carried to promote the growth of desirable trees already in existence. Though looks paradoxical, useful forests are created and maintained chiefly by removing certain category of trees and promoting certain trees judiciously.

Choice of Silvicultural System

Every forest requires a silvicultural system to rejuvenate itself. The system helps the forests to maintain its diversity, composition, stocking and productivity in perpetuity. If a forest is to continue, then it needs to be regenerated and tended, otherwise, it will vanish. Therefore, while formulating management plan, it is the duty of the forest planners to propose appropriate silvicultural system to regenerate and tend the forests on a sustainable basis. In fact, there are many silvicultural systems, which are in practice in different parts of our country. Therefore, it is imperative to choose the suitable silvicultural systems for a forest crop of any locality.

A good silvicultural system is a long-term programme of treatment designed to fit a specific set of circumstances. Silvicultural systems have been suggested for the forests of the study area after analyzing the natural and socio-economic factors of the district. A host of solutions has been proposed to go as far as possible in capitalizing on the opportunities and overcoming the difficulties found in the study area. Various considerations that have a bearing on their development in the area were weighted. While selecting the suitable silvicultural systems for the forests many concerns have been expressed and a whole range of silvicultural options have been weighted. Few of such consideration are presented below:

State of Conditions Required for Regeneration

While deciding about the system, the state of conditions that is fundamental to rejuvenation of forest crop was assessed. The points considered were light requirement of the crop and availability of seedlings in the forest floor. Consequent to this, the ease of regeneration in different forest areas was considered. Over the years, the ground conditions required for natural regeneration in the forest of the study area have been found to be steadily deteriorating. Frequent fire incidences and high grazing pressures have alarmingly marred the prospect of natural regeneration in forest adjoining to villages. This has also triggered the process of forest fragmentation in the study area. In such situations, two options are left. They are: the first is to make an effort to make the forest floor conditions compatible with requirement of natural regeneration and second is to prevent fire and to control growing anthropogenic pressures on the forests.

Efficient Use of Growing Space and Site Productivity

Silviculture is a practice of growing space management in forests. Managing growing space regulates the site productivity and stand composition. Therefore, forest growth and site productivity is a function of growing space. In other words, silviculture is a tool for enhancing forest growth and productivity. Efficient use of growing space with the help of silviculture determines the productivity of the site. While proposing silvicultural system for the study area, the type of tending operations, which will promote stand development and productivity by using the growing space efficiently, has been actively and critically considered.

Species Composition

Determining the appropriate composition of stand is a daunting task. In forests where the management objective veers round timber production, management deliberately favours certain timber species over others. Consequently, the linkage the non-timber tree provides to the rest of the community gradually gets weakened. Therefore, it was necessary to weigh these options critically as well as carefully, before proposing any silvicultural treatments for the crop.

The arrangement of compositional pattern of an ecosystem along the gradient shows the effects of past management on composition. This gradient is a tool for quantifying the degree and frequency of compositional changes that have occurred within each type of ecosystem in a landscape (Hunter Jr Malcom L., 2001). This gradient reflects the range of potential composition for that ecosystem. Total absence of natural forest or compositional pattern like in a plantation indicates one end of this gradient. In the other end, there is compositional pattern almost identical to past pre harvested period. This is the result of no human impact on the forests. Therefore looking at ecosystems along gradients of disturbance, compositional pattern can be an important conceptual tool for assessing the desirability of a particular composition in a landscape. It also provides a framework for quantifying the potential trade-offs that exists between maintaining compositional pattern and producing desired quantities of tangible products.

The management goal in the study area was to maintain compositional pattern similar to the past. This was a difficult task. But with the careful inventory analysis and appropriate tending operation this became easy. The writer could plan to arrange the current composition of stands within each ecosystem type along a disturbance gradient.

Successional Stages

Silviculture is a tool to manage successional pathways. It results in change in species composition. Sometimes, change in compositional pattern is less conspicuous. In mixed multiaged forests, selection harvesting removes a particular species or group of species on preferential basis. If small openings are made, shade tolerant species increase. If big openings are made then light demander species may increase. Therefore, based on these principles, treatment has been proposed.

Sustainable Production of Goods and Services

The principal goal of silvicultural management is to make each forest the source of indefinitely sustained and uniform flow of wood and other benefits. The difficulty of attaining this goal is greater when the production cycle of the benefits or the age of the stand on which it is dependant is long. Sustained yield is most difficult to achieve with timber and other benefits available only from stands that have ages measured in decades or centuries. However preoccupation with sustained yield for timber has not obscured the fact that the appropriate distribution of age classes also provides for a continuing flow of other benefits. It has also brought about high diversity in those ecological factors governed by stand age. Some of the benefits associated with multiple uses of forests require some stands older than optimum rotation age for timber. This consideration has also been taken care of. In general it can be said that age-class distribution that is best for sustained yield of timber has been found good for the sustained flow of all other important benefits of the forest too.

Topography and Soil Conditions

Topography plays an important role in crop composition of the forests. Bamboo forests are found in hilly, rocky and exposed mountainous area of the study area. Teak and mixed forests are also found in hilly ranges. Likewise, soil conditions also play a pivotal role in crop composition. All these factors were studied in their perspective before choosing silvicultural system for forests of any working circle.

Compatibility with the External Factors of Locality

While selecting the silvicultural system, the compatibility of the system with external factors of locality like frost, wind, draught, and proneness for pest and insect damage was also examined. Many pockets in the forest are found prone to frost. Therefore, it is necessary to adopt appropriate silvicultural system to provide protection to the seedling crop from adverse effects of locality factors.

Objects of Management

Objects of management determine the selection of silvicultural system. If the objective of forest management is to produce big timbers, than the high forest system will be the choice. Considered in this context, a large chunk of area has been put under selection-cum-improvement system. When the object is to produce small timber, pole and fuel for the local communities, then the coppice system or improvement system is generally adopted. In order to achieve the goal, certain pockets were also put under the improvement system. But this system includes prescription of coppice system as well as improvement. In all such situations, other factors were also considered while selecting the system. When the primary aim of forest management is to conserve the bio-diversity and to produce timber as a subsidiary output, then the selection-cum-improvement system with focus on conservation of bio-diversity was adopted. In a hilly and mountainous forest area, the goal of management was to protect the watershed too. Therefore, in such condition maintenance of existing forest cover remains an important aim of forest management. Apart from this, the following guidelines were also followed :

- Conservation of biological diversity in forest landscapes should be promoted by maintaining a broad range of age classes including a significant portion of old stands.
- Forest should include a mosaic of all normal stages in community development.
- Elements of spatial pattern which includes amount, portion, size, inter-path distance, and variation in patch size and landscape connectivity (Harrison and Fahrig, 1995), should be incorporated.
- Landscapes managed with uniform cutting unit sizes and spatial distribution can develop into relatively homogeneous mosaics despite apparent diversity in forest conditions. In other words, pattern diversity can be low which would result in loss of some species and processes.
- Different patterns of cutting can result in very different proportion of edge and interior forest conditions.

Intensity of Silviculture

Intensity of silviculture is the amount of effort expended on the treatment and care of forest stands. It depends on the nature and quantum of objectives. The degree of intensity is estimated in terms of such things like the frequency and severity of felling during the rotation. The appropriate intensity of silviculture varies with site quality, management objectives, market demand and nature of organization. The proper level of silviculture for each stand should be selected because the application of a common degree of intensity will not give the optimum results throughout a given forests, unless the crop is small and uniform. In fact, the forests in a landscape unit are variable and diverse. Therefore, in the context of study area, an appropriate intensity has been suggested while formulating management prescriptions.

System of Forest Management

As regards management, the Forest Management Unit (FMU) is divided into working circle or zones on the basis of management objectives. The nature of resource potentiality and policy guidelines determine the management objectives. If a productive forest consists of a stand pre-dominatingly mature, then all such compartments having such type of forests are grouped together to form one working circle or zone. To manage the stand of this working circle, a suitable silvicultural system is proposed. Detailed prescription for treatment under this silvicultural system shall be developed to form a definite silvicultural regime for that circle or zone. Likewise, if a group of compartments contains predominantly young or middle agreed crop or both, then they are bundled to form a different working circle or zone. The management objectives for this working circle will be to improve the crop, in quality and quantity. Then an accessory silvicultural system like "improvement" needs to be selected. Hence, grouping of compartment can be continued as long as all the forest area are concerned. After grouping of compartments under different working circles, based on forest condition, composition, erosion, proneness, cultural importance, protection need, the area under working circle can be further divided into sub-zones or treatment types. In other words, under each working-circle, there will be multiple sub-zones or treatment types, completely based on the ground conditions and management needs. Generally, areas of reasonably similar crops are put together to form a working circle where a common treatment can be applied (Bebarta, K.C., 2003). In certain circumstances, the working circle may overlap. As an example, the bamboo forests generally constitute an overlapping working circles.

After framing working circle and sub-types, it is necessary to decide about felling cycle. Felling cycle has been defined as the time that elapses between successive main felling on the same area (Anon, 1966). Its determination depends in the silvicultural needs of the crop as well as the objectives of management. If the objective is to produce timber of large size, then a longer felling cycle is opted. A smaller felling cycle produces small timber (Bebarta, K.C., 2003). In the context of the study area, the length of rotation, which is about 120 years, has been divided into six parts, which constitute the felling cycle. This felling cycle synchronizes with the time required for the target timber species to move from one girth class to the other. A felling cycle smaller than the proposed twenty years shall produce poles and medium sized timbers. Therefore, the proposed felling cycle stands appropriate in the context of management objective.

Under each silvicultural system, there are three important operations such as regeneration, tending and removal of selected mature trees. The method of regeneration may be of seedling or of coppice origin. Tending operations help in development of particular type of stand. The removal of mature crop consists of successive felling or selection felling. In the forests of the study area, principally two types of silvicultural systems have been adopted. They are : (1) selection-cum-improvement system, and (2) improvement felling with dominantly coppice system of regeneration. Selection system is a silvicultural programme, which aims at creating or maintaining balanced distribution of all age and size classes in a unit area. The felling is distributed spatially over all most all the area. Improvement felling with dominantly coppice system of regeneration aims at improving the crop and promoting vegetative form of reproduction. Felling under this system is also distributed spatially all over working circle area.

Silvicultral Management Options

Drawing sustenance from the National Forest Policy, 1988, the working plan of the study area aims at conserving soil and water, improving growing stock, maintaining crop composition and productivity (through natural and artificial regeneration), conserving bio-diversity along with meeting local needs of the people. Special emphasis has been given to people's participation in different facets of forest management.

Presently, the system of managing forests through working circles is in vogue. The whole area of the forest division or Forest Management Unit (FMU) was divided into different working circles or zones to meet different objective of management. Again, to meet the special ecological, social and management need, area under each compartment has been again divided into different treatment parcels or areas known as 'treatment type'. Such system of zonation and sub-zonation is a healthy management practice being practiced since long.

Constitution of Working Circles

In the forests of the study area, principally two silvicultural systems have been adopted. They are : (1) selection-cum-improvement system and (2) improvement working with coppice regeneration system. Selection system is a silvicultural programme, which aims at creating or maintaining balanced distribution of all age and size classes in a unit area. The felling is distributed spatially over all most all the area. Improvement working with coppice regeneration system aims at improving the growing stock as well as promoting vegetative form of reproduction. Felling under this system is also distributed spatially all over working circle area. The crop dominantly coppice in nature will be tended suitably along with promoting seedling regeneration all over the forest of the working circle. A separate working circle has been constituted for tending and maintaining the old plantation areas. Apart from this, separate and distinct bamboo working circles, one for traditional bamboo working and another for rehabilitation of degraded bamboo forests have been formed. Substantial bamboo areas have been allotted to these two working circles.

Different working circle has different management objectives. A list of management objectives in different working circle is presented in Table 5.1. The silvicultural system and the area allotted to different working circles are provided in Table 5.2. About 87 per cent of the study areas have been allotted to both Selection-cum-Improvement (SCI) and Improvement Working Systems. About 13 per cent of the area has been allotted rehabilitation. A map showing allotment of compartments among the working circles and felling series have been presented in Fig. 5.1 and Fig. 5.2. A detailed discussion on different working circle has been given in the next paragraph.

Table 5.1 : Management Objectives in Different Working Circles

S.No.	*Working Circles*	*Management Objective*
1.	SCI	Selection forests, high forests and crop improvement
2.	Improvement Working	Crop improvement, promotion of regeneration, micro-habitat (treatment subtype) development
3.	Restoration of Degraded Forest (RDF)	Rehabilitation of degraded forests
4.	Bamboo (Overlapping)	Development of bamboo forests and micro-habitat (treatment subtype) development
5.	Rehabilitation of Bamboo Forests (Bamboo RDF)	Rehabilitation of degraded bamboo forests
6	Plantation Management (Overlapping)	Tending of old plantations

Table 5.2 : Working Circle Area and Silvicultural System

S.No.	*Working Circle*	*Forest area allotted (in hect)*	*Silvicultural System/ Tending Operation*
1.	Selection-cum-Improvement	1.11776 (49%)	Selection and Improvement system
2.	Improvement Working	85209 (38%)	Improvement system along with coppice regeneration
3.	Rehabilitation of degraded forests	29144 (13%)	Artificial Regeneration / coppice regeneration and improvement system
4.	Plantation management (Overlapping)	11266 (27%)	Thinning
5.	Bamboo (Overlaping)	61499 (27%)	Culm selection system
6.	Rehabilitation of degraded bamboo forests (Overlapping)	1887 (1%)	Artificial regeneration of bamboo and tending operations

Note : The total forest area of the study area is 226129 hect.

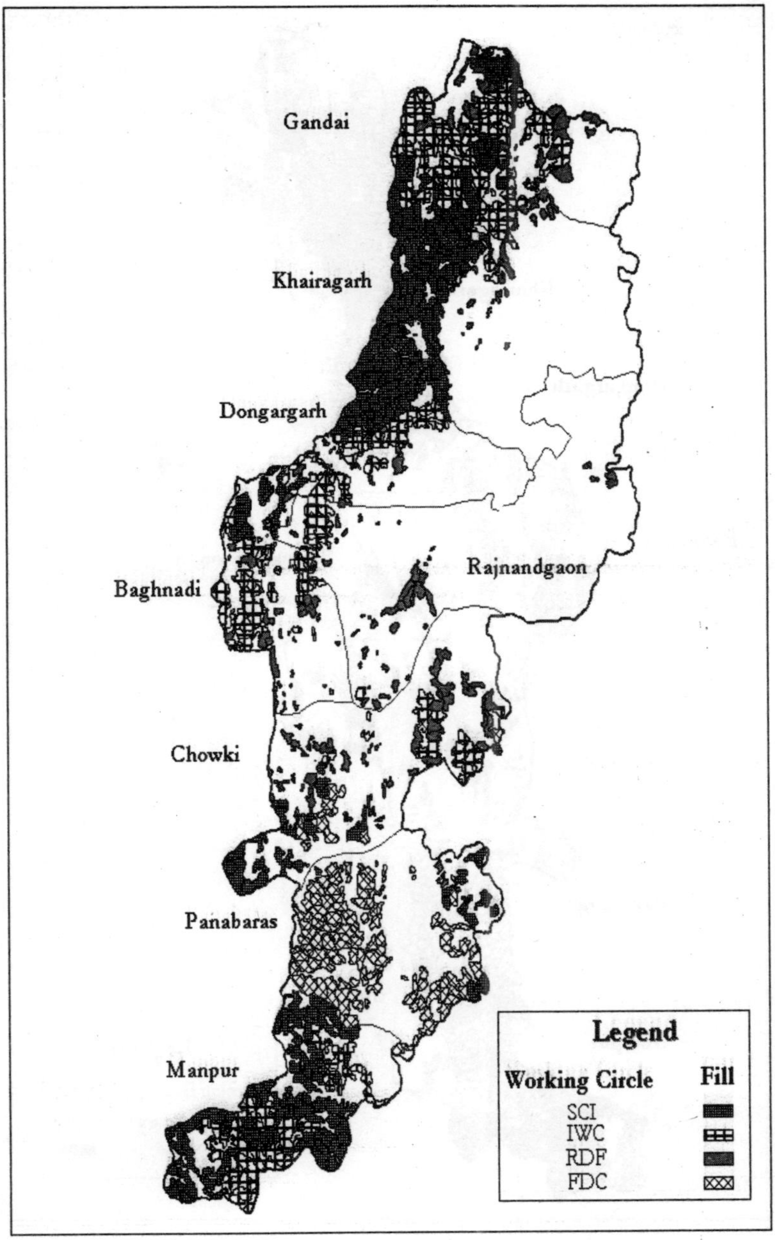

Fig 5.1 : Map Showing Forest Areas allotted to different W/C

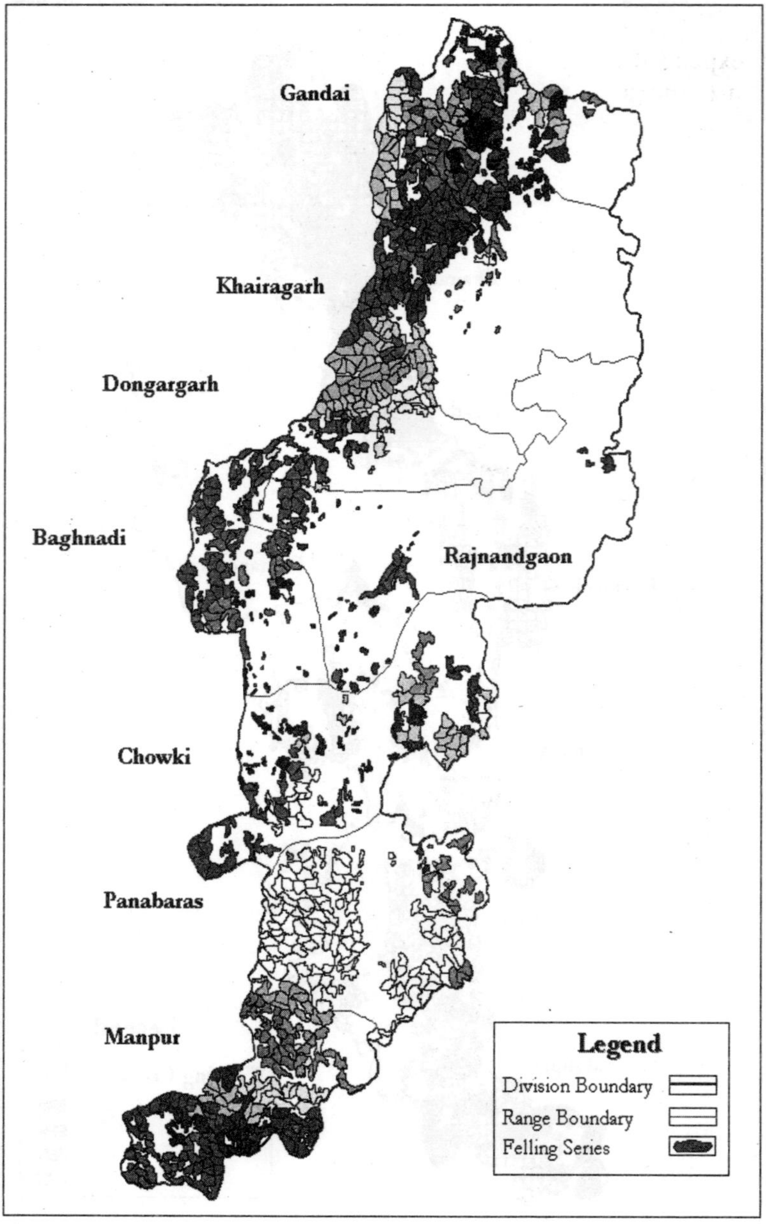

Fig 5.2 : Map Showing Felling Series

Selection-cum-Improvement (SCI) Working Circle

Forests under this working circle consist of teak and mixed crop. It also encompasses the majority of bamboo areas of the division. Forests are spread in Gandai, Kheiragarh, Dongargarh, Baghnadi, Chowki, and Manpur ranges of the study areas. The working circle encompasses not only steep, and mountainous but also plain areas. There are four principal management objectives of this working circle. They are: (1) production of large sized timber, (2) full protection of the area located on steep and mountainous part (from the point of soil and water conservation), (3) augmentation of natural regeneration primarily of seed origin, (4) crop improvement. As already said, selection system of silviculture has been adopted for this working circle. A map showing allotment of compartments into the working circle is presented in Fig 5.3.

The said system is a silvicultural programme which aims at creating or maintaining balanced distribution of all age and size classes in a unit area. The felling is distributed over almost all the area. In this system the old trees are cut in each felling cycle. Re-generation is supposed to come up naturally in the gaps created by removal of mature trees. In practice annual working on whole of areas is not practicable; this is normally because of problems such as: (1) difficulty in working over a large area, (2) higher logging and extraction costs due to low annual cut per unit area, (3) lack of balanced distribution of all age classes, and (4) lack of good seed year every year. In order to obviate some of these problems a felling cycle of 20 years has been adopted. The area of working circle has been divided into 23 felling series. Every year a single coupe from each felling series will be worked. The whole forest will be treated once during a felling cycle. That means the revisit period for every unit of the forests is 20 years.

Felling in such selection-cum-improvement system involves : (1) the removal of all dead and dying trees, (2) felling of such trees which are diseased, defective or lacking vigour, (3) removal of individuals of inferior or gregarious species to promote the individuals of valuable or desirable species, (4) removal of selection trees which are silviculturally available for felling. Different exploitable diameter or girth for different species are fixed and all such trees are felled as selection trees unless they are required to perform other silvicultural functions. A profile of crop and management characteristics of the circle is presented in Table 5.3.

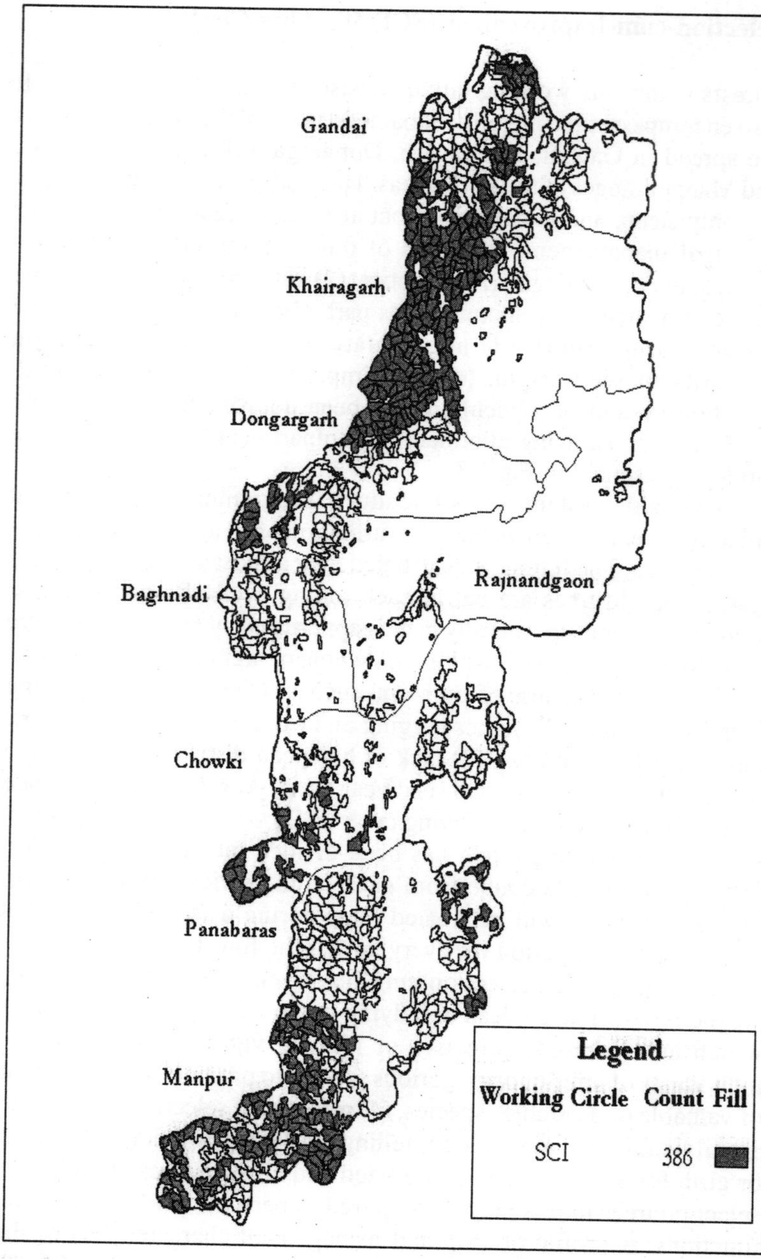

Fig 5.3 : Map Showing Compartment allotted to SCI W/C

Table 5.3 : Crop and Management Characteristics : SCI working Circle

S.No.	*Item*	*Description*
1.	Area	1,11,775.597 hectares
2.	Crop density	0.5 to 0.7
3.	Crop age	Middle aged to mature
4.	Growing Stock per hectare	74.11 volume in cmt.
5.	No. of Trees	308.410 trees per hectare
6.	Established regeneration	1136 seedlings per hectare
7.	Tree Diversity	44 (No. of tree species per hectare)
8.	No. of Felling Series	23 Nos.
9.	Method of Regeneration	Natural, coppice and artificial
10.	Felling Cycle	20 years
11.	Yield Regulation	Area and number of trees of selection girth
12.	Selection girth for Teak, Saja, Dhawada, Bija	120 cm. over bark
13.	Selection girth for Tinsa, Karra, Lendia, Bhirra	60 cm. over bark

Treatment

In order to maintain the species diversity in the area, trees of those species that constitutes more than 1 per cent of the total prevailing population in the respective working circle have been fixed for felling, provided those trees are silviculturally available. In pockets specified for bio-diversity conservation, only dead trees will be removed, leaving a wide scope for forest management with a focus on conservation of biological diversity. However, certain sang trees will be left in the forests for the promotion of wildlife. Soil and moisture conservation work will be undertaken according to field requirement and budget availability. All fruit bearing species and trees of Kullu, Kusum, Salai will be reserved against felling. Apart from this, no tree of Sisham, will be marked, since the population of this tree has declined considerably. A treatment map showing all the treatment types will be prepared for the annual felling series.

Specific treatment

The coupe area has been divided into three treatment types. They are: (1) Protection area, (2) Old plantation areas, (3) Dense workable areas.

Treatment Type "A" : Protection Areas

It includes protection areas where no felling will be done and soil and water conservation measures will be adopted. The hilly mountainous areas having slope more than 30 per cent has been put under this treatment type. This apart, all erosion prone areas have been put under this treatment type too.

Treatment Type "B" : Old Plantation Areas

It includes old plantation areas, which will be treated as per prescriptions given in plantation management (overlapping) working circle.

Treatment Type "C" : Dense Workable Areas

It includes the balance well-stocked workable area. Treatment regarding selection-cum-improvement felling will be carried out. In this working circle, Cut Back Operation (CBO) after the main felling, 6th year cleaning and 11th year thinning have been prescribed. Fire protection and grazing control methods with participatory approach have been suggested to protect the forests.

Improvement Working Circle (IWC)

The main objectives of the working circle are to protect the area and to improve the crop along with meeting the local needs. It covers medium quality miscellaneous forests of Gandai, Khairagarh, Dongargarh, Baghnadi, Rajnandgaon, Chowki, Panabaras and Manpur range. A map showing allotment of compartments into the working circle is presented in Fig 5.4.

The silvicultural system adopted in this working circle is improvement felling system or improvement system in general. In fact it is not a silvicultural system. It aims at neither regenerating a crop nor producing a crop of distinctive characteristics. It is an accessory silvicultural system. Placing the forests under such system provides breathing time for eventual management of such forests under a selection system in later years. This accessory system is a method of treatment involving essentially the removal of inferiors growing stock in the interest of better growth of more valuable individuals. When improvement felling system has been adopted in better crop areas, it has been combined with selection fellings. In such a situation, the areas have been put in Selection-cum-Improvement system. In such a treatment system, in addition to the operations carried out under improvement feelings, some sound and mature trees above the

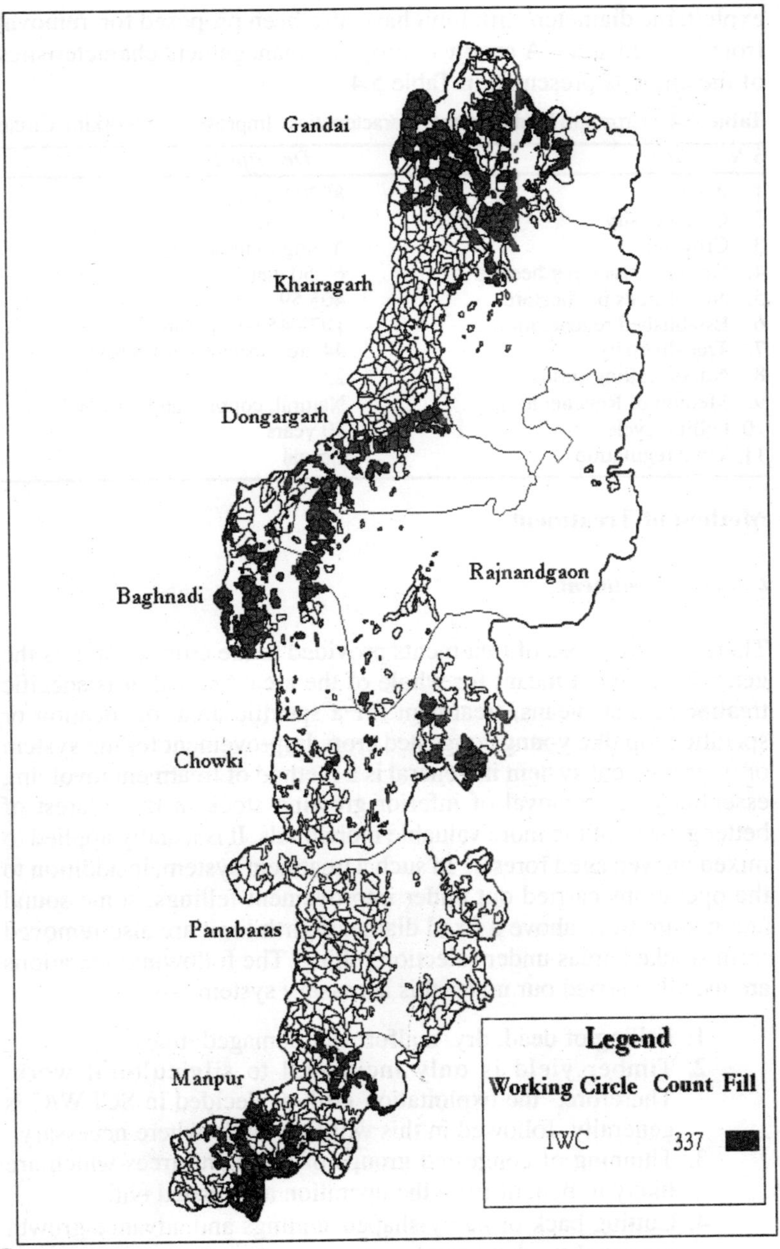

Fig 5.4 : Map Showing Compartment allotted to IWC W/C

exploitable diameter/girth limit have also been proposed for removal from stocked areas. A profile of crop and management characteristics of the circle is presented in Table 5.4.

Table 5.4 : Crop and Management Characteristics : Improvement Working Circle

S.No.	*Item*	*Description*
1.	Area	85209 hectares
2.	Crop density	0.4 to 0.6
3.	Crop age	Young to middle aged
4.	Growing stock per hectare	67.50 cmt.
5.	No. of trees per hectare	405.59
6.	Established regeneration	1079.45 per hectare
7.	Tree diversity	44 tree species per hactare
8.	No. of felling series	23
9.	Method of Regeneration	Natural, coppice and artificial
10.	Felling cycle	40 years
11.	Yield regulation	Casual

Method of Treatment

General Treatment

There are two types of treatments provided to the crop. First, it is the general treatment meant for whole of the area. Second, it is specific treatment, that means, treatment for a specific area, or location or specific crop like young congested crop. Improvement felling system or improvement system in general is a method of treatment involving essentially the removal of inferior growing stock in the interest of better growth of the more valuable individuals. It is usually applied to mixed uneven aged forests. In such a treatment system, in addition to the operations carried out under improvement fellings, some sound and mature trees above a fixed diameter/girth limit are also removed from stocked areas under selection felling. The following operations are usually carried our under this accessory system :—

1. Felling of dead, dry, malformed, damaged trees.
2. Timber yield is only incidental to silvicultural work. Therefore, the exploitation girth as decided in SCI W/C is generally followed in this working circle where necessary.
3. Thinning of congested groups of poles and trees which are likely to benefit from the operation are carried out.
4. Cutting back of badly-shaped saplings and advance growth expected to give better result in the form of coppice shoots are undertaken.

5. Removal of undesirable undergrowth of trees of inferior species, which are preventing, or likely to prevent, development of regeneration of desired species.
6. Soil and moisture conservation work are carried out as per field requirement and budget availability.
7. Planting in under-stocked and blank forest areas will be carried out.
8. Treatment map will be prepared for each annual felling coupes.
9. Trees like Kullu, Kusum, Salai and Sisham and fruit bearing trees are not marked for felling.
10. Climber cutting will be done where necessary.

Specific Treatment

The areas under this working circle have been divided into four treatment type areas on the basis of treatment required for the crop. These treatment types are :

1. Treatment Type "A" (Protection area)
2. Treatment Type "B" (Rehabilitation area)
3. Treatment Type "C" (Old plantation area)
4. Treatment Type "D" (Dense workable area)

Treatment Type "A" (Protection Area)

Protection of forest and improvement of the regenerative capacity of forest floor are the primary objectives of forest management. Forest areas situated in hilly terrain and slope having more than 30 degree are put in this treatment type areas. The crown density of the forest crop in this working circle ranges from 0.4 to 0.7. There will be no felling in such areas except removal of dead and seriously diseased trees. Soil and water conservation works will be undertaken as and when necessary.

Treatment Type "B" (Rehabilitation Area)

Rehabilitation of degraded forest is one of the primary objectives of forest management. The working circle contains forest area which is in different stages of degradation. Forests under this working circle are situated in hilly terrains as well as plain areas near habitation. The crown density of the forest crop in this treatment type ranges from 0.2 to 0.4. There will be no felling of green trees. Only dead trees will be felled. The available root stocks will be protected and promoted. Artificial regeneration such as planting will also be taken up. Soil and water conservation activities will be carried out to promote and expedite the process of ecological recovery. However, the *prime focus* of management will be on protection against cattle and fire.

Treatment Type "C" (Old Plantation Area)

Integrated management of old forest plantation in accordance with the prescriptions of a separate comprehensive plantation management working circle is the objective. This is done by excluding the old plantation areas falling in other working circle and putting these areas under plantation management (overlapping) working circle areas. It covers forest area where various plantations have been taken in the past. Basically teak and bamboo plantations have been taken up in the study area in large scale.

Treatment Type "D" (Dense Workable Forest Area)

Natural regeneration of the forests, especially of high forests and improvement of the standing crop are the primary objectives of forest management. The treatment type includes workable dense forest area where selection felling will be carried out along with light improvement felling. The crown density of such treatment type areas ranges from 0.5 to 0.6. No fruit bearing trees will be taken up for felling. In addition to it, there will be restriction on felling of those trees whose population is below 1 per cent of the total population. In this working circle provision has been made for Cut Back Operation (CBO) in the year following the main felling year. Fire protection and grazing control methods with participatory approach have been suggested to protect the forests.

Rehabilitation of Degraded Forest (RDF) Working Circle

Degradation has become a common phenomenon. Because of increasing anthropogenic pressures, some forest areas are degraded annually. In forest parlance, degradation means the phenomena of decreasing trees population in such a way that the crown density falls below 0.4. The remedy is rehabilitation of such degraded landscape.

Forest areas in ranges such as Gandai, Baghnadi, Rajnandgaon, Chowki and Dogargarh have been extensively found degraded. Therefore, these areas have been allotted to this working circle. A map showing allotment of compartments into the working circle is presented in Fig 5.5.

Rehabilitation is a planned process that aims to regain ecological integrity and enhance human well-being in [on] deforested or degraded forest landscapes. From the ecological management point of view, rehabilitation has the following components.

- *Reclamation :* Reclamation is meant to recover productivity (but little of the original biodiversity) at a heavily degraded site. In time the protective function and many of the original ecosystem services may be re-established. Reclamation is often done with native species.
- *Rehabilitation* : Rehabilitation is meant to re-establish the productivity of a degraded forest and some, but not all, of the plant and animal species thought to be originally present at the site.
- *Ecological restoration*: Ecological restoration is meant to reestablish the presumed structure, productivity and species diversity of the forest originally present on the site.

From the point of silvicultural management, rehabilitation involves : (1) a package of solutions, such as natural regeneration, enrichment planting, selective logging, re-introduction of native species, protecting soil and water resources, (2) connecting forest fragments and creating a link between protected areas and well-managed forests, and (3) valuing forest goods and services to begin to quantify and evaluate how stakeholders can directly benefit from these.

The objective of this working circle is to rehabilitate the degraded forests of the division. In includes mostly under stocked, degraded and blank forest areas found scattered all over the study areas. The density of such forests ranges up to 0.4. Generally such areas are found near the habitation where the anthropogenic forces are very severe. This apart, the factors of locality in such areas are also very poor. A profile of crop and management characteristics of the circle is presented in Table 5.5.

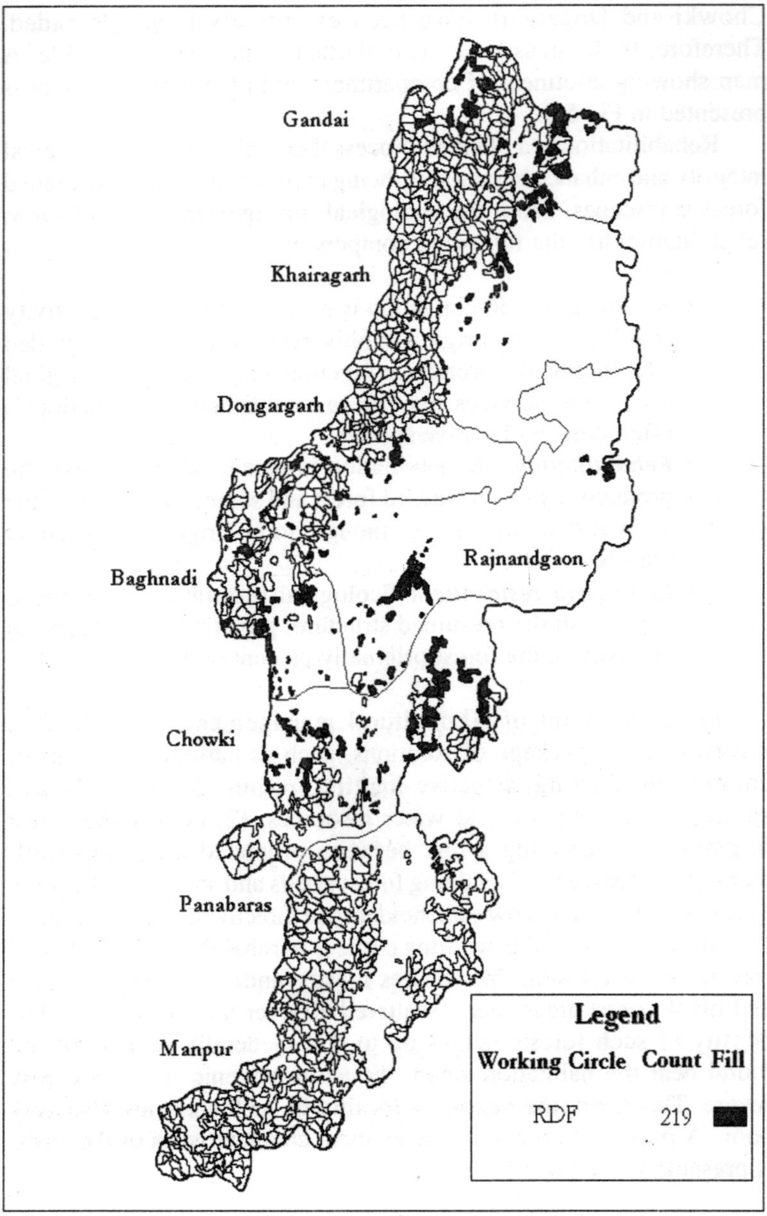

Fig. 5.5 : Map Showing Compartment allotted to RDF Working Circle

Table 5.5 : Crop and Management Characteristics: Rehabilitation of Degraded Forests Working Circle

S.No.	*Item*	*Description*
1.	Area	29,143.522 hectares
2.	Crown density	< 0.4
3.	Mean volume per hectare	36.42 cmt.
4.	No. of trees per hectare	213.7
5.	Established regeneration	654 Seedlings per hectare
6.	Tree diversity	27 Tree species per hectare
7.	No. of treatment series	18
8.	Method of regeneration	Natural, coppice and artificial
9.	Treatment cycle	10 years

General Treatment

The general method of treatment in these working circle areas is to protect and to regenerate the area by natural as well as artificial regeneration. Wherever, there are good patches of forest stand found, treatment of crop improvement shall be applied. In the working circle rehabilitation of degraded forests will be undertaken. There are four treatment types.

Specific Treatment

Treatment Type "A" (Protection Area)

It includes protection areas. Such areas will be put under rigid protection regime. The focus is on protection from fire, illicit grazing and illicit felling. Soil and water conservation measures will be undertaken as and when necessary. The forest area suffering from erosion proneness has been identified and action plan prepared.

Treatment Type "B" (Rehabilitation Area)

It includes forest areas, generally having sufficient rootstock with potentiality for recovery. Suitable plantation and silvipasture development work will be undertaken along with rigid protection regime.

Treatment Type "C" (Old Plantation Area)

In includes old plantations, which will be treated according to prescription under the plantation management (overlapping) working circle.

Treatment Type "D" (Dense Forest Area)

It includes balance workable dense forest areas. However, in such areas only dead trees will be removed. Fire protection and grazing control measures have been prescribed. With respect to management of such forests with people's participation, village forest committees have been constituted. In such case, management of such forests is done with the participation of local joint forest management committees.

Plantation Management (Overlapping) Working Circle

Constitution of this working circle aims at providing focus on management of old plantations for tending and other subsidiary silvicultural operations. At present the study area have 11265.650 hectares of old plantation. (Table 5.6) The working circle includes all successful teak, bamboo and miscellaneous plantations carried out in the past years. These plantations are spread over ranges of Gandai, Khairagarh, Rajnandgaon, Dongargarh, Baghnadi, Chowki, Panabaras and Manpur. Chowki and Gandai range have the highest teak plantations. The first plantation of teak in the study area was raised in the year 1955. Mixed plantation has been raised since 1972. The species planted in such mixed plantations are siris, eucalyptus, amla, neem, karanj, khamar and bamboo. However, such mixed plantations have never been successful. Pure bamboo plantations have also been raised since the year 1974 in three ranges such as Gandai, Chowki, and Manpur.

Table 5.6 : Extent of Different Plantations

S.No.	*Plantation*	*No. Compartments*	*Area in hectares*
1.	Teak	87	6759.54
2.	Mixed	44	3938.00
3.	Bamboo	10	568.08
	Total	141	11265.65

Detailed treatments with respect to cleaning, thinning and other subsidiary silvicultural operations have been prescribed. With respect of teak plantation, clearing at 5th year and thinning at 11th, 21st year, 31st year, 41st year have been prescribed. Looking at the lack of uniformity in growth, all thinning have been proposed to be silvicultural thinning. Thinning regime and schedule for the entire list of plantations has been prepared and added in the working planformulated by the author. In the past working plan, many teak plantations due for thinning were not taken up. Such plantations have been identified and grouped for arrear thinning taking into account of their congestion and potential scope for growth. In the mixed plantation only 5th year cleaning has been proposed. In addition to all these, it has been proposed to protect all the plantations from fire and grazing incidences with a participatory approach.

Rehabilitation of Degraded Bamboo Forest Working Circle

The study area has 1887.285 hectares of degraded bamboo forests. These bamboo forests have been degraded in the last plan period. These areas have been classified as degraded bamboo forests because they contain 31 to 50 clumps of bamboo per hectare. Such forests are found in the understorey of miscellaneous forests. They are distributed in ranges such as: Gandai, Khairagarh, Baghnadi, Dongargarh and Manpur. Manpur range has degraded bamboo forest of about 800 hectares. The primary objective behind constituting this working circle is to provide focus and priority to rehabilitation of degraded bamboo forests. The working circle has three distinct management objectives. They are :

- To improve the existing bamboo clumps and promote clump formation in areas adjacent to old clumps afresh.
- To augment the natural regeneration of bamboo.
- To undertake under planting of bamboo in suitable areas.

The first objective can be achieved by improving condition in and around bamboo clumps. The second objective can be achieved by improving the physical conditions in the forest floor. This will rejuvenate the forest floor ecologically. This is possible if we take fire and grazing protection measures properly. The third objective can be achieved by undertaking bamboo plantation in open forests. Since the area is small, a single treatment series has been proposed. An account of crop and management characteristics of the forests is presented in Table 5.7.

Table 5.7 : Crop and Management Characteristics: RDF Bamboo Working Circle

S.No.	*Item*	*Description*
1.	Area	1887.285 hectare (17 compartments)
2.	Crop Density	Very low density
3.	Site Quality	2nd and 3rd site quality
4.	Clumps per hectare	31 to 50
5.	No. of Trees per hectare	213.7
6.	Established Regeneration	654 seedlings per hectare per hectare
7.	Tree Diversity	27 tree species per hectare
8.	No. of Treatment Series	1
9.	Method of Regeneration	Natural, and artificial
10.	Treatment Cycle	10 years

Treatment Method

Important rehabilitation measures such as improving the site, protection of bamboo areas, stimulating natural regeneration and artificial planting will be taken under this working circle.

In this working circle Rehabilitation of degraded bamboo forests will be undertaken by identifying two treatment types. They are:

Treatment Type "A" (Natural Bamboo Patch)

Areas falling under this category will be taken up for improving the site. This will promote conservation and development of (bamboo plantation area) of bamboo clumps.

Treatment Type "B" (Open Forests)

The area under this category will be taken up for bamboo planting. Such areas are generally forests found near village boundaries.

Bamboo (Overlapping) Working Circle

The bamboo forest of the study area extends up to 61499 hectares. This apart, an area of 1887 hectares has been categorized as degraded bamboo forests. There degraded bamboo forests have been dealt separately in RDBF working circle. The objective of bamboo overlapping working circle is to conserve and use the bamboo resources of the forests sustainably. The working circle areas include all the good bamboo areas found in Gandai, Khairagarh, Dongargarh, Chowki, Panabaras, Manpur and Baghnadi ranges. The bamboo species is *Dendrocalamus strictus.* The proportion of dense, medium and low dense bamboo areas is 36 per cent, 51 per cent and 13 per cent respectively. Very good bamboo forest is found in ranges like Khairagarh, Dongargarh and Gandai. Out of these, the first two ranges have very good bamboo forest comparable to any international standard. The two ranges of Donargarh and Khairagarh have hilly terrains with step slopes perfectly suitable for bamboo forests. A map showing allotment of compartments into the working circle is presented in Fig 5.6.

The total growing stock of bamboo forests is 856 lakh culms. It constitutes about 1,78,333 metric ton. Out of it, the first year culm constitutes 138.85 lakh culms (16%), the second year 239.60 lakh culms (28 %), and third/fourth year 477.57 lakh culms (56 %). This is an asynchronous distribution of bamboo culms between 1st year, 2nd year and 3rd/4th year bamboo culms. Ideally the proportion should have been 25 per cent, 25 per cent and 50 per cent. It means, the first year culm population is about 9 per cent below the desirable limit. In other words, one can safely deduce that the rate of recruitment of bamboo is not satisfactory and it speaks volume about the unsustainable management of bamboo forests.

Bamboo working has been going on since the beginning of last centuary. But the silvicultural system under which it is worked has not been properly named. Since the bamboo fellings are done on culm selection basis, in such a way that the production of new culms take place continually. Therefore, the name could have been culm selection system. Bamboo is being worked on the basis of a cutting cycle of four years.

Management plan prepared for managing the forest resources of the study area includes prescriptions and treatment for scientific harvesting and development of bamboo forests. As a rule, good bamboo forests are harvested in a cycle of four years. It means, in every fourth year felling of bamboos are carried out. In such operations, generally speaking, only the third and fourth year bamboo culms are removed. It is because of the fact that the longevity of a bamboo culm

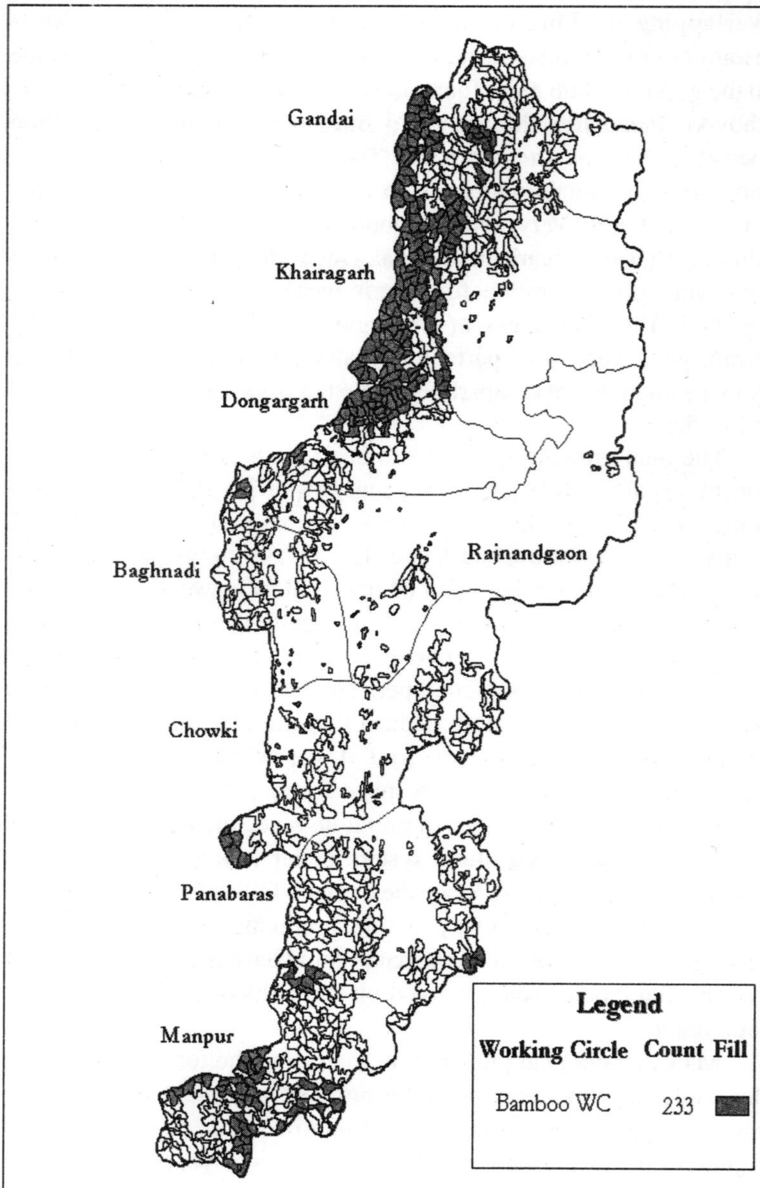

Fig. 5.6 : Map Showing Compartment allotted to Bamboo W/C

is not more than five years. If such old and matured culms are not removed then they die and get dried up, which consequently increases the proneness of bamboo forests to fire. Therefore, sustainable management of bamboo forests necessitates periodical harvesting and development including protection of such resources. The crop and management characteristics of bamboo working circle is presented in Table 5.8.

Table 5.8 : Crop and Management Characteristics: Bamboo (OL) Working Circle

S.No.	*Item*	*Description*
1.	Area	61499.291 hectares
2.	Crop Density	Dense and medium dense
	I. Dense Bamboo Forests	33977.258 hectares
	II. Medium Dense Forests	18523.483 hectares
	III. Low Dense Forests	8998.55 hectares
3.	Growing Stock	856 lakh culms (178333 NT)
4.	Clumps and culms per hectare	87 clumps and 1862 culms
5.	No. of bamboo clumps per hectare	70 (Average)
6.	No. of Treatment Series	54
7	Method of regeneration	Natural and artificial
8.	Cutting cycle	4 years

Treatment Method

Bamboo areas have been divided into two treatment types. They are: treatment types 'A' area and treatment type 'B' area. Treatment type 'A 'area has dense bamboo forests with healthy bamboo clumps and treatment type 'B' area has very low density with malformed bamboo clumps. Commercial bamboo cutting will be taken up in accordance with standard bamboo felling rules in all those treatment type 'A' area only. Rehabilitative measures will be taken in treatment type 'B' area.

Standard bamboo felling rules have been prescribed. Salient features of treatment area :

- Bamboo will be worked on the basis of a cutting cycle of four years.
- Restriction in cutting of one and two year old culm.
- Retention of older bamboo for support of immature culms.
- Retention of minimum number of bamboo based on the clump size.
- Prohibition of digging of rhizomes.
- Insistence on cutting of bamboo with a sharp instrument.
- Bamboo cutting will be carried out when the seed has been

completely shed in case of flowering. The period of harvesting is November to end of March.

Treatment Type

The treatment types have their own management objectives. On this basis, the treatment strategy has been formulated. The areas have been divided into two main treatment types. They are :

Treatment Type 'A' (Dense Bamboo Forests)

Areas falling under this category will be taken up for commercial cutting of bamboo culms on selection basis. This operation will be carried out along with subsidiary silvicultural operations.

Treatment Type 'B' (Degraded Bamboo Forests)

Areas falling under this category will not be taken up for commercial cutting of bamboo culms. Rather they will be taken up for rehabilitating the site. This will promote conservation and development of bamboo. This operation will be carried out along with subsidiary silvicultural operations. The area under this category will also be taken up for bamboo planting, if found suitable.

Estimated Yield of Bamboo

The total area under bamboo forest is 61499.291 hectares. From the point of site quality of bamboo forest, the distribution among 1st, 2nd and 3rd site quality is 36 per cent, 51 per cent and 13 per cent respectively. This indicates that majority of the bamboo area belongs to site quality II and above. In other words, it symbolizes availability of potential bamboo forests. The growing stock of bamboo is 856.02 lakh notional ton (Table 5.9).

Table 5.9 : Growing Stock of Bamboo

S.No.	*Culm Type*	*Number of bamboo (in lakhs)*	*Per cent*
1.	1st year culm	138.85	169
2.	2nd year culm	239.60	28
3.	3rd year culm	477.57	56
	Total	856.02	

The pattern of bamboo cutting is cyclic in nature. In every fourth year, the bamboo forest is visited for cutting. Cutting takes place as per the standard norms of bamboo cutting rules. Every year 15374 hectares of bamboo forest is available for bamboo cutting. The stock available for cutting every year is 214 lakh culms (Table 5.10).

Table 5.10 : Annual Stock of Bamboo Available for Cutting

S.No. Bamboo Culm	*No of Culms (in lakhs)*
1. 1st year culm	34.71
2. 2nd year	239.60
3. 3rd year	119.39
Total	214.0

The average bamboo clumps available in the bamboo forest is 81 per hectare. Since 50 per cent of the bamboo forest being of II quality, according to bamboo cutting rules at least 15 culms per clump should be left standing. On this basis, the following number of bamboo has to be left standing in the forests.

R = No. of bamboo culms to be retained.
A = Annual bamboo area taken up for cutting.
C = Clumps per hectares.
N = No. of minimum culms per clumps to be retained.

$$R = A \times C \times N$$
$$= 15374 \times 81 \times 15$$
$$186794 \text{ or } R = 187 \text{ lakh Culms.}$$

Therefore, the number of bamboo which will be available for cutting will be : Annual Stock – No. of bamboo culms to be retained. Calculating this, we find : 214 – 187 = 27 lakh Culms. Taking the average length of bamboo 5 metres, there will be 6000 NT of bamboo available for cutting. In the year 1980, Forest Survey of India (FSI, 1980), in its inventory report, cautioned against the decreasing availability of green bamboo in the bamboo forest of the study area. Therefore, looking into the condition of bamboo forest and the scope for their development, we can safely estimate that the yield will be about 6000 NT of commercial bamboo and 10 to 12,000 NT of industrial bamboo, *i.e.* roughly two times of the prospective commercial yield.

Flowering Behaviour of Bamboo

Bamboo differs significantly from other vegetation in its flowering behaviour. Generally, most bamboo species flower gregariously at fixed intervals and all culms including those of current year as well as rhizome die after flowering. *Dendrocalamus strictus,* the most common bamboo, is known to flower both sporadically and gregariously at long intervals of 20–65 years. In case of sporadic flowering, only the clumps that flower die and not all the clumps occurring in the entire area.

Gregarious flowering in *Dendrocalamus strictus* is a recognised conspicuous phenomenon. The characteristics of gregarious flowering are: (i) flowering occurs almost in the entire area, (ii) it involves almost all or some portion of the clump, (iii) flowering takes place in all the culms in a clump, (iv) flowering is followed by death of the clump, (v) it follows a cycle which occurs after a long interval, (vi) it takes few years, commonly 2 to 4 years to complete flowering in the area. The period between two gregarious flowerings over the same area is believed to be somewhat constant and is called physiological cycle. Different periods for physiological cycle have been reported from different areas. With respect to Chhatisgarh some records of gregarious flowering are available. In bamboo forests of South Bastar region, the first reported occurrence of gregarious flowering was in the year 1948 and the second occurrence was recorded in the year 1981. In Bilaspur the first and second gregarious flowering was recorded in the years 1895 and 1960 respectively. In Kawardha bamboo forests, the first and second gregarious flowering took place in the years 1910 and 1954 respectively. Likewise, the first gregarious flowering of bamboo in Raipur took place in 1924 and the second in 1960 after a gap of 36 years. Looking at these data, it appears that the bamboo forests of the study area, more particularly Khairagarh and Dongargarh ranges, may witness gregarious flowering at any time in the first decade of this century.

Regulation and Management of Miscellaneous Subjects

No silvicultural management will be complete without miscellaneous regulation of subjects or activities not compatible with forestry. All silvicultural programmes will achieve their objective, if forces, which are inimical to forestry, are curbed in time. Therefore, it is imperative to identify those forces and to draw ways and means to minimize them or their effect on forests. The forces may be promotional in nature, but its threshold has acquired a proportion, which is inimical to elements of conservation. Take for example the case of rights and concession. This promotes sustainable use of forest resources but once the limit of use is crossed, it starts to leave an adverse impact on the resources base itself. Likewise, there are certain practices which are altogether destructive to forest recourse base. Take for example illegal cultivation inside forests or encroachment in forest land.

Therefore, other than the technical aspects mentioned in working circles, there are many subjects, which have a bearing on sustainable management of forest resources. Such subjects are required to be regulated and managed. Otherwise, in spite of our best of efforts in silviculture domain, we may not achieve our management objective. Few of such miscellaneous subjects are elucidated below.

Rights and Concessions

Rights and concessions are some of the important social instruments by which community can be associated in forest management process. Therefore, forests of the state recognize traditional rights and concessions. This means_entry into forests and use of the produce thereof, by the people living in and around the forests is free. The provision of Nistar remains a main plank in forest management in the study area. The domestic requirements of fuel wood, fodder, minor forest produce and construction timber have been increasing, leaving a negative impact on the forest processes. The moot point, however, is to regulate and control such instruments to ensure sustainability of forests. Nistar will be given as per annual Nistar Patrika. Non-nationalized forest produce will be allowed free to the local people. The basic regulation regarding regulation and management of Rights and Concessions have been prescribed in the plan document.

Management of Ecotone Forests

Bamboo forests constitute an important component of the forest ecosystem in the state. The study area has at least two pockets of ecotone areas requiring special management practices. Such forests are not only ecologically important but also socially important because they provide basic needs that constitute the essential element of the livelihood security of the poor people in the study area. Such forests are found in the northern boarder, from where the sal forests begin and in south where the miscellaneous forests meet the teak forests of Maharashtra and Bastar.

Conservation of MFP Including Medicinal Plants

Non-wood forest produces like Tendu Patta, Chironji, Kullu, Honey, Harra, Baheda, etc. constitute a major plank of livelihood of the tribal and poor people of the district. NWFPs provide a range of goods for domestic use and for sale in the market. They are a source of emergency sustenance in times of hardship – when crop fails, and economic crisis occurs. They are also a primary source of income to the poor. There is a need for taking appropriate initiative for *in situ* and *ex situ* conservation, domestication and non-destructive harvesting with active support of local people.

Forest Protection

Forest, being an open access resource (OAR) are vulnerable to various kinds of negative practices such as theft, illegal grazing and encroachment. Protecting forests holds the key to ecological sustainability. The plan document lays sufficient emphasis on protection mechanism.

Diversion of Forest Land

The Forest Conservation Act, 1980 and the National as well as State Forest Policy vow for total regulation of forest land. Policy enunciates that diversion of forest land for any non-forest purpose should be subjected to the careful examination by specialists from the standpoint of ecological, environmental and social costs and benefits. This has been properly emphasized in the plan document.

Wildlife Conservation

Conservation of wildlife and bio-diversity is one of the principal objectives of management. In the study area, over the years, lot of changes has occurred in the habitat of wildlife. Provisions have been made to improve the availability of water, shelter and the habitat as a whole. The count of big wild animals is presented in Table 5.11. There has not been any substantial increase in population of big game animal.

Table 5.11 : Wild Animals in the Study Area

S. No.	*Name of Wild Animals*	*No. of Animals*		
		Year 1996	*Year 1997*	*Year 1998*
1.	Tiger	9	9	9
2.	Panther	62	66	66
3.	Cheetal	250	249	249
4.	Sambhar	35	35	35
5.	Nilgai	23	25	25
6.	Bear	34	36	36
7.	Hyena	695	700	700
8.	Wild boar	456	450	450
9.	Fox (*siyar*)	848	856	856
10.	Gaur	6	8	8

It has been found that there are four conservation pockets in the study areas, which are rich in tree diversity (Table 5.12). These pockets have been identified on the grounds. In order to conserve the bio-diversity and habitat for wild flora and fauna of these pockets, special rules have been prescribed in the plan document. The main objective behind curving out certain ecologically rich pockets is to provide full protection from ill effects of timber harvesting operations. Beside this, other rules necessary for conservation as a whole have been formulated and proposed for their practice. Tree diversity index for these four patches have been calculated. The treatment has been designed by keeping in mind that with minimum interference from man the forces of ecosystem will pre-eminently dominate the process of ecological succession. This will help to keep the 'flux of nature' at minimum.

Table 5.12 : Conservation Area

S.No.	*Range*	*No. of Compartment*	*Area in hectare*
1	Gandai	7	2450.145
2	Dongargarh	9	2721.50
3	Chowki	6	2473.53
4	Manpur	8	2431.490
	Total	30	10076.215

Treatments suggested for those conservation pockets are as follows :

- Full protection against fire and grazing will be given.
- Fruits and seeds from the trees will not be collected.
- No grass will be collected.
- No improvement felling operation will be carried out.
- Only mature trees and dead tress will be marked for felling.

Participatory Forest Management

Joint Forest Management strategy for protection and development of forests areas through participation and co-operation of local people have been discussed in detail in chapter six. The basic approach and action plan for participation in forest management has been given in the plan document.

Forests Extension and Social Forestry

To inculcate people's participation in forest management and plantation inside and outside forests, extension programmes have been outlined in the plan document. They include individual and community programmes that encourage sustainable use of forest resources including the use of non-conventional source of energy and wood alternatives. Training of staff has also been emphasized in the plan document to make the programmes successful.

Other Miscellaneous Regulations

Maintenance of Forest Boundaries and Pillars

5-year demarcation scheme has been drawn in the plan document. A broad scheme has also been prepared to get the forest pillars repaired on a programmatic basis.

Grazing Control

Grazing is proposed to be regulated as per provisions of Madhya Pradesh Grazing Rules, 1986 and amendments made from time to time. Grazing restrictions for specific area have been prescribed in the plan document.

Fire Protection

All fire cases will be properly recorded. Strict fire protection will be observed. Training of personnel in modern fire fighting methods has been suggested. A broad scheme elaborating fire protection has been prepared and presented along with the plan document.

Forest Villages

There are five-forest villages in the study area. Villages will be managed in accordance with forest village rules and other government instruction provided in the plan document.

Sample Plots

There is no sample plot in plan area. It is proposed to layout couples of growth plots in teak as well as mixed forest patches of Khairagarh, Dongargarh and Manpur ranges. Beside this, a couple of sample plots is required to be laid in proposed conservation areas too.

Marking

Detailed procedure for marking of trees has been drawn in the plan document. The staffs engaged in executing demarcation and marking of trees are expected to follow it in letter and spirit.

Logging and Felling

Felling and logging methods and procedures have been suggested in the plan document. The field staff are expected to follow the methods and procedures in letter and spirit.

Irregular Exploitation

Any felling outside the annual treatment series will be considered as an irregular exploitation for which sanction of competent authority will be essential. In such cases prior permission is required from the competent authority. In such situation as a safeguard, quantity equivalent to that felled under irregular felling will be retained in the annual coupes.

Soil Conservation

Appropriate soil and water conservation measures have been prescribed to tackle problem area in eroded areas. Details of soil and water conservation techniques are discussed in the plan document.

Protected Tree

Procedure for declaring protected trees has been given in the plan document. Such trees will be protected from felling.

Pest and Insect Control

Various measures for insect and pest control have been incorporated in the plan document. In case of any incidences of pest and insect attack on forest crop, the procedures should be followed.

Maintenance of Maps

Procedures regarding maintenance of maps such as stock map, management map topographic maps and other thematic maps have been laid out in the plan document.

Control and Records

For exercising effective control in the field and over the implementation of plan prescription, maintenance of standard records have been prescribed in the plan document.

Establishment and Labour

Training and welfare activities for staff and labour have been suggested in the plan document. Apart from this, the present staff strength needs improvement. The concerned divisional forest officers of the divisions shall undertake an appraisal exercise and draw a reorganization plan if necessary.

6

Mainstreaming Participatory Forest Management

Managing forests on the basis of sustainability principle transcends beyond the domain of technical forestry. It involves myriad issues like social, economic, political betterment of human condition, social equity, poverty reduction, gender roles, governance structure, etc. Such issues lie at the heart of sustainable development and use of forest resources. Addressing these issues requires the participation of a broad spectrum of people in the process of forest management.

This chapter deals with different issues and case studies related to participation of people in forest management. It is presented in four sections. First section introduces the subject of participation with a focus on concept, definition and relevance. Second section deals with a discussion on practice of participation in the state. The third section dwells upon two specific context driven case studies. The first case study was conducted in the year 1998 in three village level JFM committees. The purpose of the study was to assess the adaptation of people in fuelwood stress condition. The study tried to determine the degree of fuelwood stress in these villages and it's impact on participatory management also. The second case study pertains to the appraisal of performance of Joint Forest Management (JFM) activities in the study area as a whole. The evaluation of working of JFM committees in the study area was done with an innovative field inspection based and computer aided grading methodology. The fourth section deals with the attendant problems in operationalising joint forest management and difficulties in achieving participation in a hierarchical organization like forest department and the need to change the working for scaling up community participation in forest department.

I

The concept of people's participation is an integral part of our philosophy of democratic welfare state. To start with people's participation in actual practice is little more than getting contribution from them, in terms of labour, cash, gifts of land and materials for the works programme. In such practices the idea of promoting self-reliance among people is normally not pre-dominant. This is simply participation in the work without any involvement. Unless there is involvement in the process of work, the very purpose of participation is defeated. It is only by getting people involved in the process of improving themselves through their own efforts that we can develop their capacity, self-reliance and cohesion in the community.

Community participation is gaining acceptance among policy makers in India as an essential aspect of managing natural resources. This belated recognition is reflected in the Government of India (GOI) initiatives to facilitate participation of community in management of forests. It has received considerable attention because it is central to any strategy to improve natural resource management. JFM was initiated as a sequel to policy imperatives of National Forest Policy, 1988. The first notification regarding the guidelines for initiating participatory mode of management came in the year 1990. Afterwards, in many international conventions the focus remained on participation of local people in management of natural resources. Almost all countries adopted people's participation as the means of developing and managing the precious natural resources. Forest department of all states initiated participatory mode of management of forests with the grand programme of joint forest management. The new approach, being the antithesis of the classical model, faced host of opposition from the forest departments. However, understanding the changing paradigm of forest management all over the globe, the forest departments of all the states began a comprehensive programme to put JFM in practice. The first step towards this was laid with the process of institutionalizing JFM in all states. Institutionalizing JFM being a social and political process, takes a long time, normally at least 20 to 30 years, to show results. Therefore, it is not opportune time to assess the performance of the forest department. At best forest department's performance in the field of institutionalizing JFM can only be assessed. Viewed from this perspective, performance of many state forest departments has been commendable. A good indicator of the performance of the forest departments is the stabilization of area under dense forest cover to around 38 million hectares during the last decade.

People's participation has acquired special significance in the development of a society in the last two decades. It was acknowledged that majority of people in India live below the level of basic requirement, hence the quest for a new approaches to development has dawned in the minds of planners. Worldwide, the concept of endogenous development is gaining momentum. It centres on man and underlines the importance of participation of local people in the development of natural resources. Even, FAO in 1979 has agreed by consensus that "participation was basic human right and may be regarded as an ethical aim if we are to attain participatory democracy." Effective participation assumes significance when it becomes inherent into the social life of the people. Thereby, it becomes part of the institutional system. Management of forests by people's participation shall herald a new beginning in the social milieu the rural society.

Forest resources, all over the country are under stupendous pressure. Anthropogenic factors with varying intensities have started degrading these resources. This has created a situation of grave scarcity of renewable forest resources in most of the parts of the country. Most of our forests have lost the regenerative capacity. However, with the introduction of participatory approach of forest management, a ray of hope to revive the ailing forests has appeared in the horizon. There are encouraging reports indicating people's participation in forest and water resources from different parts of the country. In Madhya Pradesh, also earnest effort has been taken for participatory management of forests. However, success is not found in every such effort. Some good efforts have been taken in Rajnandgaon forest division too. Though, the success seemed to have taken the shape of "Island" in a land of failures, it is high time to ponder over the reason for success in such micro-level and to find ways and means to replicate such changes in other forest areas.

II

Participation : Concept, Definition and its Relevance

The development literature is rich with definitions of the term participation. Participation means different things to different people.

Definition of participation depends upon the context and background in which participation is applied (Box 6.1). Since it is not possible to review all those definitions in this essay, this discussion shall confine itself to some of the most important ones. According to UNEP (1991) 'participation means that people are closely involved in the economic, social, cultural and political processes that affect their lives'. FAO (1982) emphasizes that participation in all decision-making processes but by the people's own organization and through self-organized action. Paul (1987) introduces the concept of enhancement of well-being in terms of values cherished by the communities as being the ultimate objective of participation. This appears to be the most simple definitions of participation since it implies that people participate only in those economic, social, cultural and political processes, that have the potential to bring about favourable changes in their lives. This definition comes very close to the notion of Johnston and Clark who have strongly warned against regarding participation as a 'free good, desirable in unlimited quantities.' Participation, according to them, "has a large cost for the poor and they will invest their participation when they believe it will secure them valuable benefits not otherwise available at comparable cost, time and risk".

A review of the literature on participation as well as the ways in which participation is operationalized in different development interventions reveals that participation is conceptualized and understood differently. The manner in which participation can be enlisted also varies. Various attempts have been made to develop a typology of participation (Arnstein, 1969; Hart,1992; Pretty, 1994). Box 6.2 gives details of seven types of participation as conceived by Pretty *et al.* (1995). On this basis participation can be viewed along a spectrum with passive participation at one end and self-realization at the other end. Passive participation is where people are told to do what to do. On the other end is self-mobilization where the people themselves are in total command. As one moves from passive participation to self-determination, the control of the local people and outsiders over the processes varies. At one end of the spectrum, say in the case of passive participation, people's control is almost non-existent while at the other end, in case of self mobilization, people have almost total control over the processes while the role of outsiders is at minimum.

Community participation is also defined as people acting in-groups to influence the direction and outcome of development programmes

Box 6.1 : *Defining Participation*

- Participation is defined as a voluntary contribution by the people in one or another of the public programmes supposed to contribute national development, but the people are not expected to take part in shaping the programme or in criticizing its contents (Economic Commission for Latin America, 1973).
- Participation includes people's involvement in decision making processes, in implementing programmes, their sharing in benefits of development programmes and their involvement in efforts to evaluate such programmes (Cohen and Uphoff, 1977).
- What gives real meaning to popular participation is the collective effort by the people concerned to pool their efforts and whatever other resources they decide to pool together, to attain objectives they set for themselves. In this regard participation is viewed as an active process by which the participants take initiatives and actions that are stimulated by their own thinking and by deliberations over which they exert effective control. The idea of passive participation which only involves the people in actions that have been thought out or designed by others and controlled by others is unacceptable (ACC Task Force and Working Group on Rural Development Programme Harmonization, Rome, 1978).
- People's participation is essentially to do with economic and political relationship within the wider society; it is not just a matter of involvement in project activities but rather the process by which rural people are able to organize themselves and, through their own organization, are able to identify their own needs, share in design, implement, evaluate participatory action (FAO, 1982).
- Community participation is an active process by which beneficiary or client groups influence the direction and execution of a development project with a view to enhancing their well-being in terms of income, personal growth, self-reliance, or other values they cherish (Paul, 1987).

that affect them (Paul, 1987). The key aspects of this definition are people acting collectively and influencing the outcomes. Collective action is necessary, as many aspects of forest development require coordination between member of the communities as well as neighbors. Their influencing outcome is the core of participation. Participation at a basic level is to share and to own. It is a process of mutual adjustment and orientation of behaviour among participants. Mutual adjustment requires communication and interdependence as

Box 6.2 : *Typology of Participation*

- *Passive participation*: People participate by being told what is going to happen or has already happened. It is a unilateral announcement by an administrator or project management without listening to people's responses. The information being shared belongs only to external professionals.
- *Participation in information giving*: People participate by answering questions posted by extractive researchers using questionnaire surveys or similar approaches. People do not have the opportunity to influence proceedings, as the findings of the research are neither shared not checked for accuracy.
- *Participation by consultation*: People participate by being consulted, and external people listen to views. The external professionals define both problems and solutions, and may modify these in the light of people's responses. Such a consultative process does not concede any share in decision making, and professionals are under no obligation to take on board people's views.
- *Participation for material incentives*: People participate by providing resources, for example labours, in return for food, cash, or other material incentives. Much of the on-farm research falls in this category, as farmers provide the fields but are not involved in the experimentation or the process of learning. It is very common to see this kind of participation, yet people have no stake in prolonging activities when the incentives end.
- *Functional participation*: People participate by forming groups to meet predetermined objectives related to the project, which can involve the development or promotion of externally initiated social organization. Such involvement does not tend to occur at the early state of project cycles or planning, but rather after major decisions have been made. These institutions tend to be dependent on external initiators and facilitators, but may become self-dependant.
- *Interactive participation*: People participate in joint analysis, development of action plans, formation or strengthening of local institutions. Participation is seen as right not just the means to achieve project goals.
- *Self-mobilization*: People participate by taking initiatives independent of external institutions to change system. They develop contacts with external institutions for resources and technical advice they need, but retain control over how resources are used. Such self-initiated mobilization and collective action may or may not challenge exiting inequitable distributions of wealth and power.

Source : Pretty(1994) in pretty *et al.* (1995).

action of one depend on the actions of another. Therefore, participation is a voluntary process, which may take place in several stages beginning with appreciation in which various parties listen to each other, share perceptions and validate each other's information.

There are other scholars who have tried to interpret participation in slightly different ways by placing more emphasis on the process rather than on the contents and outcome of the participation. In this regard special reference may be made of Fraser, a social scientist, who makes a distinction between participation that is spontaneous or bottom-up with the participation that is induced, coerced, or top-down participation. Similarly, the United Nations (1993) distinguishes between coerced participation, which it condemns, induced participation, which it regards as second best, and spontaneous or bottom-up participation, which it contends, 'comes closest to the ideal mode of participation, as it reflects voluntary and autonomous action on the part of the people to organize and deal with their problems unaided by governments or other external agencies.' Further, Midgley, another social scientist also makes a similar distinction between 'authentic' participation, in which local people democratically control project decision-making and pseudo-participation, in which projects are carried out according to prior decisions made by outsiders.

The academicians have also tried to equate participation with achieving power: that is power in terms of access to, and control of, the resources necessary to protect livelihood. The following are some of the statements illustrative of this understanding of participation :

- .... The promotion of popular participation implies a redistribution of power and this calls for scientific analysis which give due recognition to political factors, social forces and the role of social change;
- ....Participation is concerned with distribution of power in society, for it is power which enables groups to determine which needs, and whose needs, will be meant thorough the distribution of resources; and
- .... Power is the central theme of participation and participatory social action entails widely shared, collective power by those who are considered beneficiaries. The people become agents of social action and mere differentials between those who control, and need resources is reduced through participation.

Participation in forest management requires, at a minimum, a executing agency, which not only follows a blue-print for development, but involves the communities in analyzing conservation problems and identifying strategies to tackle them. Other than the executing agency, three aspects are critical and constitute the driving force to such a process. The first aspect is the ability of members to participate in a community action or to have a collective voice or decision made jointly by the community and implementing agencies. The second aspect is community organizations. Whenever needed, implementing agencies need to work with communities to initiate collective process. This process, which is generally referred as social organization, is critical in forest management. It is because it distributes benefits and costs evenly or on the basis of parity and calls for shared sacrifices while managing forests. If mechanisms are not put in place to ensure that Joint Forest Management is in everyone's best interests, those who bear the costs may undermine development efforts and investment on such projects will be wasted. Therefore, a mechanism to translate development needs into action at ground level needs to be developed jointly by communities and executing agencies. The third aspect is the willingness of the community to share a part of the costs. This is also important because it increases the degree and quality of participation in community action. These three key aspects of participation have implications for each other and overall level of community participation in forest management.

Participation in Practice

We examine the practice of participation in terms of the four key elements identified: social organization, empowerment, joint decision-making, cost and benefit sharing. These key elements influence the process of joint forest management and determine success in most cases. Before examining these aspects, it is relevant to discuss the spread of JFM committees in the study area so that participation can be put in proper context.

The district of Rajnandgaon constitutes the study area. It encompasses two territorial forest divisions: (1) Rajnandgaon forest division and (2) Khairagarh forest division. Earlier both the divisions were in one unit called Rajnandgaon forest division. Bifurcation took place in the year 2000. Two types of committees have been formed in both the divisions. One is Village Forest Committee (VFC) and the other is Forest Protection Committee (FPC). The forest area of both the divisions comprises about 2261 square kilometers. Presently in both the divisions, 120 FPCs and 146 VFCs have been formed. An area of 617.04 square kilometers have been allotted to such JFM committees. It means twenty seven per cent of the forest area in the study area is under joint forest management. The details of area allotted to different committees in different ranges are presented in Tables 6.1 and 6.3. The profile of members enrolled in different joint forest management committees is presented in Table 6.2.

Table 6.1 : Extent of Forest Area under Joint Forest Management

(Area in hectare)

S.No.	*Forest division*	*Area under FPC*	*Area under VFC*	*Total area*
1.	Kheiragarh	17013.462	17349.37	34362.832
2.	Rajnandgaon	20784.853	6556.42	27341.273
	Total	37798.315	23905.79	61704.105

Note : FPC denotes Forest Protection Committee, and VPC denotes Village Protection Committee.

Source : Forest Department Records: Year 2004.

Table 6.2 : Profile of JFM members

S.No.	*Forest division*	*Male-female JFM members*			
		Type of committees	*Male Members*	*Female Members*	*Total*
1.	Kheiragarh	FPC	4123	2013	6136
		VFC	6036	3900	9936
2.	Rajnandgaon	FPC	5183	2438	7621
		VFC	9229	4461	13690
	Total		24571	12812	37383

Source : Forest Department Records: Year 2004.

Table 6.3 : Area under F.P.C. & V.F.C. *(Area in hectare)*

S.No.	*Range*	*Area under VFC*	*Area under FPC*	*Total area under VFG/FPC*
	Rajnandgaon Division			
1.	Baghnadi	4044.625	1188.455	5233.08
2.	Chowki	1867.575	2733.965	4601.54
3.	Khujji	988.625	1048.57	2037.195
4.	Manpur North	3808.431	76.640	3885.071
5.	Manpur South	6871.342		6871.342
6.	Panabaras	1644.085		1644.085
7.	Rajnandgaon	1560.17	1508.79	1560.17
	Total of Division	20784.853	6556.42	27341.273

(Contd.)

Table 6.3 : *Contd..*

Kheiragarh Division			
8. Chuikhadan	2176.470	4256.690	6433.16
9. Dongargarh	882.825	–	882.825
10. Gandai	1348.597	3914.57	5263.167
11. Khairagarh	2139.04	425.720	2564.76
12. North Bortalab	3212.775	864.56	4077.335
13. Salewara	3398.925	4730.005	8128.93
14. South Bortalab	3854.83	3157.825	7012.655
Total of division	17013.462	17349.37	34362.83
Grand Total	37798.315	23905.79	61704.105

Source : Forest Department Records: Year 2004.

Social Organization

Social organization aims to organize people who can articulate their needs, participate in planning, implementation and evaluation of the programme and to develop the necessary institutions to build up community assets. A separate JFM committee for a single or a group of villages is generally required to implement the programme. Each family in the village has at least one member in those committees. The rights and duties of each member of the community with respect to the objective of the organization must be well defined and members must be educated. An action plan to carry out those organizational activities should be in place.

Before undertaking organizational activities, it is generally felt necessary to conduct PRA. The adoption of PRA methods is the first that may be taken to facilitate community participation. Though the methods are used widely today, it is primarily to rapidly gather information from communities to set up committees and obtain their approval for predetermined plans. Participatory approaches can be effective if those who initiate the exercise believe that the committees have a sound understanding of their environment and they need to be involved in finding solutions to their problems. PRA comprises of methods, behaviour and attitudes (Chambers, 1983). The limitations of the methods, evens when carried out properly are not adequately understood, as PRA is a public process, participants may initially hesitate to make interventions that may lead to open conflict; there may be appearance of consensus as a result. Communities, therefore, may need to be given time to rethink what they have suggested in initial meetings.

Approaches to Community Organization

There are three broad approaches to develop community organizations. They are : (1) building block approach, (2) direct approach, and (3) leadership-oriented approach. In the building block approach, small groups are organized for initiating larger community level organizations. The general principle in this approach is to organize homogeneous sub-populations into cohesive groups before bringing them together in a village level organization. However, the entire community may not be organized into small groups. Groups are usually initiated among the poor who are least likely to participate in village level activities. The leaders of such groups are encouraged to play an active role in village level organizations. This approach, therefore, makes it feasible to develop village organizations without depending on socio-economically powerful individuals. This approach encourages establishing what are known as 'user groups' comprising of people benefiting from project activities. The representatives of these user groups make up the executive level of the village committees. Though the user groups are small but they are qualitatively different from Self-Help-Groups (SHGs). They tend to be collection of people engaged in similar activities, where as SHGs are groups organized to help themselves through collective efforts. Presently Self Help Groups (SHG) engage in savings and lending activities in rural area. In direct approach implementing agency initiates community organizations by working through existing community organization such as Gram Panchayat or by directly establishing a dialogue with majority of people as village level. During this process, key individuals are identified and drafted in the committee. As this approach relies on existing organizations or socio-economically powerful individuals or both to develop committees, there may be limited opportunities to build organizations that have values and methods of working that are different from these of existing village organizations. The third approach is characterised as individual or leadership oriented approach. The moral authority of a person is usually critical to building the community organizations. The moral and political influence of prevailing leaders can be useful for organizing communities.

The aforesaid three broad approaches offer different possibilities of building effective organizations. The approach generally adopted is a combination of all or any of two or three. The building block approach is the most common. It takes minimum of six months to two years to establish community organizations. It requires staff willing to work silently with the communities. It offers a high probability of building effective organizations as it ensures partici-pation of socio-economically weaker sections of communities.

The leadership-oriented approach demands the least time. The resulting community organization tends to be effective but there is a likelihood that organization may fall when the leaders withdraw. This is a serious limitation of this approach. It may take a few weeks to couple of months depending on existing social capital and leadership. The effectiveness of organizations emerging from this approach is heavily dependent on the nature of existing social capital of the leader.

Assessing Effectiveness of Social Organization

Social Organization yields expected outcomes in direct relation to the efforts put into it. One criterion for assessing the capability of organizations engaged in maintaining common properties is their ability to devise and enforce rules relating to use and maintenance of common properties as village pastures and water resources. Generally, one should assess the success of social organization by measuring wide range of collective activities. However, collective action and activities require rule making and rule enforcement. Another indicator of social organization could be of Shramdan or voluntary community labour devoted to some task of common interests. This apart, there may be many criteria for assessing effectiveness of social organization.

Empowerment

Empowerment plays a critical role in participation. It is an intrinsic part of participation. Karl, a social scientists says that 'empowerment is a process and is not, therefore, something that can be given to people.' She further says that, ' the process of empowerment is both individual and collective, since it is through involvement in groups that people most often begin to develop their awareness and the ability to organize to take action and bring about change.' Karl further explained that 'empowerment is a multifaceted process, involving the pooling of resources to achieve collective strength and countervailing power, and entailing the improvement of manual and technical skills; administrative, managerial and planning capacities, and analytical reflective abilities of a local people.' Empowerment is a process geared towards participation, greater decision-making and transformative action through awareness and capacity building. In the UNDP's Human Development Report (1993:21), it has been very categorically mentioned that, 'since participation requires increased influence and control, it also demands increased empowerment. Any proposal to increase participation must, therefore, pass the empowerment test – does it increase or decrease people's power to control their lives.' To sum up, empowerment means creating awareness among people, developing ability to take collective decision and delegating authority. Involving the communities in decision making concerning whether to undertake a project with the choice of activities and technology enhances their participation. This process develops an aptitude among community members for taking right decision in right time. This also aids the process of empowerment.

Decision-making in JFM Committee

Decision-making is a key element in people's participation. It involves and commits people in development process. It begins with the decision to form JFM committees. To form a JFM committee in a village or not is an important decision for some community members. It is because JFM activities may restrict the benefits accrued from forests on a short term basis. This apart, some member may have to forgo few benefits. Those who are loosing benefits will insist on steps to ensure that they be compensated adequately. There are three types of approaches adopted in forest department (FD) to form committees. In the first approach, the FD only undertakes a consultation with local people, but without much effort to ensure that everyone has a voice or that those consulted are representative of the community as a whole. In the second approach, villagers hold a Gram Panchayat meeting to decide whether to form a JFM committee. In this approach it is difficult to ensure that the weakest members feel free to voice their opinions as the final decision to form committee rests only with the Sarpanch. In this approach decisions may be made in consultation with an entire community. This decision is based on the preferences of a specified percentage of people (either 50% or 70% of the households). Fourth, the most stringent form of community participation requires full consensus of all members of the community in the decision to form a JFM committee. Obviously, this is much more difficult task to achieve and can only work with a great effort to ensure that every one is satisfied with the fact that formation of JFM committee will be in their interests.

Choice of Activities

Choice of activities is an important facet of decision making at the committees level. The committee members should be given the powers of making decision regarding choice of activities. As communities are not given the opportunity to participate in decision-making process, priorities in the implementation deviate substantially from those of the communities. Communities place high priority on harvesting water along the drainage line in the forests to irrigate crops. Farmers are not keen on tree plantation in forest areas. They prefer to develop their private land with the help of government initiatives. Two important reasons for differences between forester's and villager's priorities are villagers preference for immediate returns and dislike for working collectively on activities that may only bring marginal returns. The poorest in communities may have to make big sacrifices by forgoing certain benefits from forests in the short run.

Likewise, decision regarding choice of technologies is also very important. It needs to be addressed properly. The concept of social fencing for protection of forests has endeared many social workers and forester. Villagers feel that the time is not ripe for translating the concept into action. Social fencing will be effective if the scale of collective activity in a village is very high. Such villages are rarely found. Therefore, there is a need for dialogue and careful targeting of technical issues. Where congruence in priorities cannot be achieved, certain trade-off may need to be made to ensure success. Effective activities not acceptable to communities will not yield benefits particularly if they require maintenance by communities. Therefore, acceptability of activities/technologies to communities is as important as their technical performance.

Cost and Benefit Sharing

Cost and benefit sharing is an important element in participation. Theoretically, it may not look important, but at the ground level, it plays a significant role in making participation work for the community development. Cost sharing is meant to ensure that stack-holders take forest development activities seriously and feel a sense of ownership on assets they create. It is practiced in some areas of Andhra Pradesh and Maharashtra. The extent and nature of cost sharing varies according to the policies or programmes of implementing organizations. In many states, a certain percentage of the cost is routinely withheld from the wages paid under different government programmes. The extent of costs shared by farmers is influenced by both villagers' willingness to pay for activities and the effort made by project agency to implement cost sharing. Villagers shall be willing to pay maximum if the activities that are being carried out are in line with the villager's individual interests.

A larger hurdle to making villagers share costs is government staff's reluctance to cost sharing. It reduces the staff's control over the forests as it requires collaboration with communities and transparent process. A key indicator of progress in the government is the disbursement of funds. Any thing that comes in the way of rapid disbursement, such as the need to consult villagers is perceived to be detrimental to performance of its duties.

Cost sharing is a better option than the input subsidy. It can improve incentives all around. But, experience of our social forestry project in Madhya Pradesh is illustrative of hollowness of claims regarding cost sharing at input level. During that phase, saplings were initially distributed to farmers free of cost from the department's nurseries. Farmers used to take more seedlings than they needed or capable of caring for. As expected, the survival rate was very poor. When farmers were required to pay for the sapling in the first phase of the World Bank Project in MP, farmers demanded better quality seedlings. Farmers did not take more saplings than they needed and the over all impact of tree planting remained favourable. Additionally, cost sharing has desirable effect on the working of implementing organizations also. It can be a potentially effective tool for making departmental staff accountable to communities. Other than cost sharing, benefit sharing is also important. At present, the benefit sharing mechanisms are not distinct. Sometimes, it appears that sharing of benefits between government and community is at the mercy of the government. Likewise, the mechanism of benefit sharing between community members is also not distinct and structured. Therefore, lack of such mechanism acts as a disincentive to people for carrying out community action.

In sum, an effective strategy for operationalising participation needs to include social organization, joint decision-making, benefit sharing and cost sharing. Social organization is a necessary condition for community participation but it is not sufficient. Sufficient conditions are created through joint decision-making in which the communities get control over the activities that affect them. Cost sharing by communities further instills sense of ownership and helps in targeting limited resources. The four aspects of participation are interrelated. Pursuing one without the others may dilute participation. The effectiveness of joint decision making in terms of meeting the community needs is influenced by the extent to which decision-making opportunities are granted to communities and the capacity of village organizations articulate their needs and engage in dialogue with the departmental staff.

III

Case Study 1 : Adaptation of People to Fuelwood Stress Situation

This study was the part of working plan revision process. It broadly aimed at obtaining details of pattern of management already in vogue and exploring scope for strengthening it. With this objective, a cluster of three optimally successful village forest committees in terms of joint forest management were taken up as sample villages for detail study. The study found out adaptation of people from three geographically separate committees to fuelwood stress situation.

Dongargaon, Chowki, Churiya, Rajnandgaon and Chuikhadan are the highly populous developmental blocks in the study area. The population density in such blocks are more than 100. Forests in those developmental blocks are degraded. However, it is heartening to note that the district has per capita forest land of 0.38 hectare against the national figure of 0.08 hectare. This situation provides enormous scope for making this district self-sufficiency in fuelwood and other minor forest produce. The favourable positioning of the district in the productive agro-ecological regions of India also provides ample opportunity for agro forestry. Considered in this backdrop, the Joint Forest Management (JFM) has great future in this district.

Objectives of the Study

The objectives of the study of the study were as follows :

1. To determine the degree of interest and involvement of people in participatory management of forests in the sample villages.

2. To study the genesis of motivation among people to protect and manage forest resources in the sample villages.
3. To study change in energy consumption pattern in sample villages.
4. To analyse the existing pattern of participatory management with its strengths and weakness, so that the *present state of joint forest management in the forest division can be improved.*

Study Area

To undertake the study, three village forest committees from Rajnandgaon forest division were selected. The basis of selection of these committees was relatively better performance in joint forest management front. For this the author carried a preliminary discussion with the officers of the division. These villages are Sangli, Maldongri and Maregaon. These village committees were found to have taken some distinctive initiatives for the conservation of forests allotted to them. All these villages are near to Chowki. Chowki is a small Tahsil township in Rajnandgaon district. All the villages are well connected by *pucca* roads. On an average, all such villages are situated about a distance of 45 kms from Rajnandgaon and 2 to 5 kms from Chowki. A map showing location of these villages is presented in Fig 6.1.

Methodology

The present study was undertaken as a part of revision of working plan of Rajnandgaon forest division. The study was conducted by using participatory rural appraisal methods. Semi-structured questionnaires were prepared to facilitate better involvement of respondents in the discussion and to obtain descriptive and qualitative data on resource availability, present state of management, community conflict and working pattern of village level forest committees. Secondary data were collected from divisional forest office and district statistical office, Rajnandgaon.

Case Studies

Case studies are presented here in the following sub-heads :

- Habitat, user group and socio-economic environment.

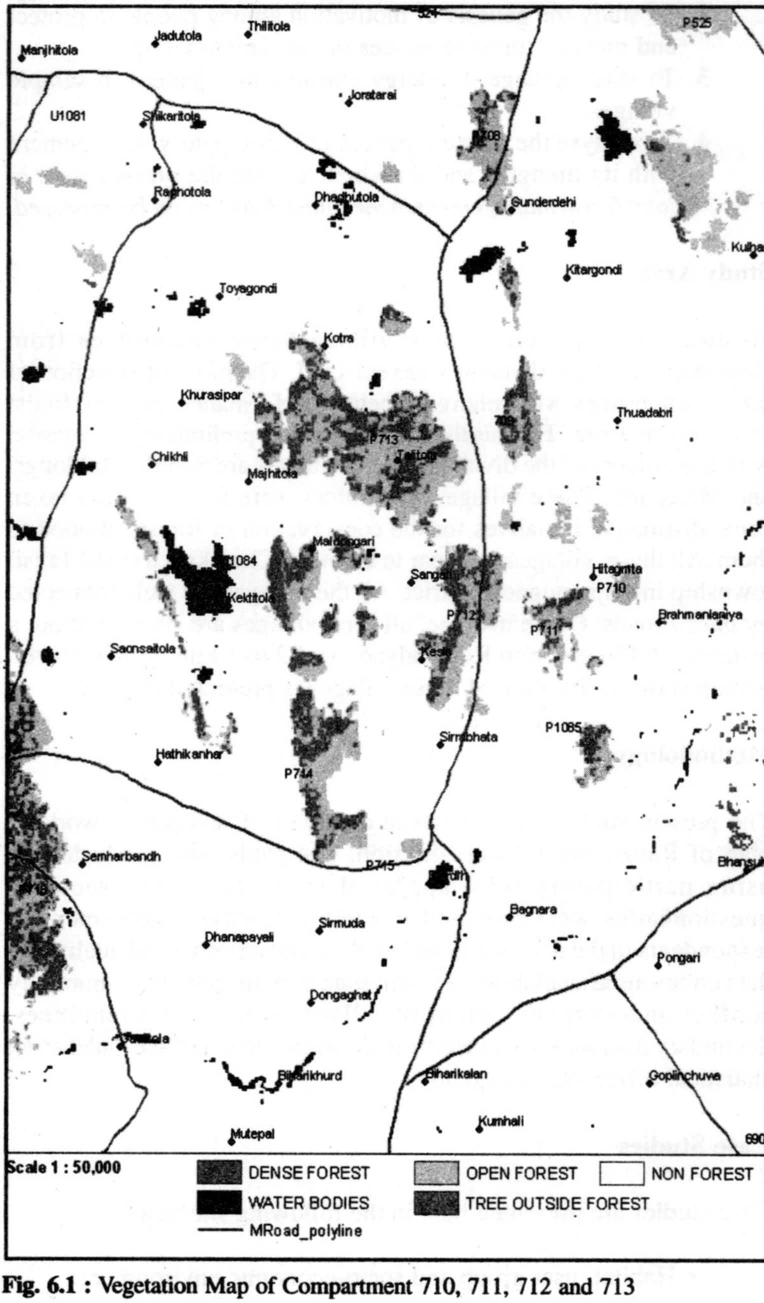

Fig. 6.1 : Vegetation Map of Compartment 710, 711, 712 and 713

- Forest resources in the study villages and their management pattern.
- Change in energy consumption pattern.
- Effectiveness of village level forest committee and their impact on forest resources.

Habitat, User Group and Socio-economic Environment

Village: Maldongri

Maldongri village is situated in Chowki forest range. It is situated along the main metal road. The total population of the village is 332. Out of it, male constitutes 103 and female 149. There are 75 households in the village. The average household size is 4.5 which ranges from 3 to 6 individuals in a household. Gond tribe is the main community in the village with sizeable presence of 63 households.

Agriculture is the major source of income. The average annual income per household from agriculture is estimated to range between Rs. 6000 to Rs. 10,000. Labour is periodic and is a supplementary source of income. It has been estimated that every household has 4.66 acre of land on an average. However, these agricultural lands are rainfed and, to some extent, remain fellow. There is 15 acres of revenue land remaining fellow. Paddy is the main crop in the village.

Village : Sangli

Sangli village is situated in Chowki forest range. It is only 2.5 kms from the main metal road. The total population of the village is 1485. The village has 275 households. Out of the total population, male constitutes 570, female 532 and children 383. The average household size is 5.4, which ranges from 4 to 7 individuals in a family. Majority of the population (i.e. 54 per cent) constitutes the general caste and rest belongs to the Gond tribals. The literacy percentage in the village is about 36 to 40 per cent.

Agriculture is the mainstay of the village economy. The average annual income per household from agriculture is estimated to be around Rs. 10,000 to Rs. 12,000. Majority of the people in the village comes in the category of marginal and small farmers. Every household has, on an average, 4.4 acres of aerable land. Majority of land is rainfed, but some lands have the facilities of irrigation on account of two irrigation ponds in the village. Labour is periodic and is a supplementary source of income. Paddy is the main agricultural crop in the village.

Village : Maregaon

Maregaon village is situated in Chowki forest range. The village boundary is only 50 metres away from Rajnandgaon-Chowki main metal road. The total population of the village is 318. Out of it male constitutes 126 and female 192. Majority of the population belongs to general class followed by Scheduled Tribe and Caste. The average size of the household is 4 to 5.

Agriculture is the mainstay of the village economy. The average annual income per household from agriculture is estimated at about Rs. 10,000 to Rs. 12,000. Majority of the people belongs to medium and small farmer category. Paddy is the main agricultural crop in the village.

Forest Resources

Village : Maldongri

These are two forest compartments (compartments 712 and 713), which are situated near the village. Both the compartments are contiguous and belong to protected forest category. The area of both compartments is 314.57 hectares. The density of the forests ranges from 0.4 to 0.5. The site quality of the forest is IVa. The major species found in the forest are Saja, Bija, Teak, Dhawda, Tendu, Kusum, Salai, Karra, Kasai and Bamboo. In the year 1985-86, teak and bamboo plantation was raised in compartment 402. However, due to lack of proper protection, it could not be successful.

In the year 1994 (11-9-94) village forest committee was formed. Now, the committee protects the forests. Before 1994, such forest lands were managed by forest department. The villagers were protecting a patch nearby consisting of revenue forests since 15 years and receiving a part of their basic requirements from that. Long association with that patch of revenue forests created awareness among the people. This awakening made the villagers to take initiative regarding participatory management of forest land. Therefore, because of public awareness and action both the forest patches are managed properly.

Village : Maregaon

The protected forest (PF) available near the village provides the basic requirements with regard to forest produce. The PF compartments are 711 and 712. The area of the compartments is 66.40 hectares. The site quality is Vb. The density of the forest ranges from 0.3 to 0.4. The major species found in the forests are Teak, Saja, Bija, Dhawda, Tendu, Kusum, Salai, Karra, Kasai and Bamboo.

Village : Sangli

The forest is confined to protected forest compartment numbers 710 and 711. Out of these two compartments, 60.03 hectares have been allotted to the Sangli Village Forest Committee (VFC). Beside this, the village has a patch of 30 hectares of good revenue forests also. The density in the forest of the PF compartments ranges from 0.3 to 0.4. There is profuse established regeneration in the forests. The main species found are Dhawda, Bija, Saja, Mahua, Karra, Achar etc. The site quality is IVb. Now, the forest is completely protected with the participation of village people.

Change in Energy Consumption Pattern

There is a lot of change seen in the energy consumption pattern of the village. Firewood is still a major fuel source in the village. However, over a span of 20 to 30 years, dependence on forests for firewood declined. This was perhaps because of scarcity of fuel wood. Scarcity of fuelwood is attributable to degradation of forests situated near habitation. Now the forest land adjoining the village does not have any dense stand of trees to cater to the fuelwood requirement. This has created a situation of resource constraint.

As a corollary to degradation, adaptation to the varying condition of fuelwood scarcity began in the seventies. The relatively well off people resorted to alternate sources of fuel like kerosene. The fact that people used electricity partially also remained as common phenomena. In the absence of trees in the forest land near village, extraction of fuelwood from trees standing in the farm became a regular practice. Another practice became prominent in the rural landscape. The agricultural wastes which were not used in the past as firewood took a center stage now. Previously, twigs and other refuse were not used as fuelwood, but now-a-days, all materials, having fuel value are used very judiciously. The "Ipomea" which emerged in India as a weed in sixties, is now used as one of the main source of fuel for the rural poor in these villages. Apart from this, there is another discerning change in the village social and cultural landscape. Due to increase in population of the village, the use of wood-dust gained momentum. Wood dusts are procured from nearby sawmills at Rajnandgaon and Chowki and are used as an potent-alternative to fuelwood. With sustained supply of wood dust, this practice is also well entrenched in the society. In the last few years, other non-conventional energy sources like gobar gas and solar cooker are also tapped but these are extremely limited. Few smokeless *"Chullahs"* are also found in the village, but they have not made any perceptible change in energy consumption pattern.

Beside this, level of dependence on fuelwood declined in such villages with the increase in their economic standard. This has enabled the people to change from the use of earthen utensils to metallic utensils for cooking food. This change in cultural habit of the people drastically reduced the energy consumption of the household. However, in spite of all such adaptability shown by the people to the situation of fuelwood scarcity, the forest resources of the nearby forest degraded primarily on account of population increase. The people in such scarcity areas have already started responding to such situation by minimizing the consumption of fuelwood and by increasing the production of fuelwood in the wasteland as well as farmland.

Effectiveness of Village Level Forest Committee and their Impact on Forests

Village forest committee were formed in the year 1994 (11-9-94) and duly registered by the divisional forest officer, Rajnandgaon forest division. The area of Protected Forests (PF) compartments such as 710, 711, 712 and 713 have been proportionately distributed to village committee of Maldongri, Sangli and Maregaon for general protection and conservation. All these compartments are degraded forests. The distribution of degraded forests is given in Table 6.4.

Table 6.4 : Distribution of degraded forest to three Village Forest Committees.

S.No.	*Name of the Village Forest Committee*	*No. of PF compartment allotted*	*Area of the compartment or part of compartment allotted*
1.	Sangli	711(Part) & 710	60.03 hectares
2.	Maldongri	712(Part) & 713	259.55 hectares
3.	Maregaon	711(Part) & 712(Part)	66.40 hectares

The cluster of villages have a history of participatory management of common resources like revenue forests and tanks. The revenue forests which are available in the proximity of the villages were protected by the people for their *bona fide* Nistar requirement since 15 years. The revenue authority over the years did not invoke any control over these forests. People in the area has a high level of awareness regarding management or common resources. As a result, gradually the people nearly increased their control and this led to proper protection and management of those revenue forests. However, in the other hand, people living in these cluster of villages felt alienated on account of increasing control of forest department over aforesaid Protected Forest. Gradually, with the increase in demand for fuel, illegal removal of fuelwood from aforesaid forest rose manifold leading to the present situation.

In the year 1994, when the idea of participatory management was suggested to the people, they agreed forthwith. The villagers have formed three villages forest committees. Such committees are engaged in protection and conservation of the forests. Now there is a profuse regeneration of important and useful species found in these compartments. Because of closure, edible grasses have also come.

Now, the villagers have a new awakening towards protection of forest. They feel that they have a right as well as a responsibility towards protection of forest. Every committee sends at least two members, one male and one female, to the forest for keeping a watch against illicit felling and grazing. The committees have established a practice of imposing fine of Rs. 30 per person who defies the decision of the committee with regard to watch and ward. There are almost no cases of defiance in any village. People collect grasses from the forest and take it to their village for feeding their cattle. Singling operation, were carried out in these forests in year 1996. The materials produced were distributed among all the members of the committee.

Present status of the forest is very encouraging. Since these forest had ample rootstock, with the protection, new coppice shoots have come up. In compartment 401, a sample plot was laid to find the status of established coppice regeneration. It was found that the forest now comprises about 7000 coppice saplings per hectare. The average girth of the sapling is 14 cms and average height is 4.5 metres. The emerging crop contains mainly, Bija, Dhawada, Karra, Landia, Saja, Amala, Mahua, and Teak. Beside this, the area has conserved enough moisture to increase the biomass.

With the help of remote sensing data, an evaluation of the impact of joint forest management on forest conservation in compartments allotted to three villages was done by the author in the year 2003. In order to do this, the forest cover data of the year 2001, obtained from Forest Survey of India, Dehradun was used. These data were generated from LISS III data of NRSA, Hyderabad. Based on this cover data, the vegetation map of those areas has been prepared (Fig 6.1). The said map indicates that almost all forest area allocated to those committees were fully regenerated and conserved. As a result of this, forest stockings in the compartments 710, 711, 712 and 713 have improved substantially.

The degraded forests allotted to three committees have been rehabilitated. This became possible because of the participatory model of forest management. However, creating a congenial atmosphere for development of such institution like "Village Committee" in those villages was a daunting task. But, with the incessant efforts and dedication of Shri L.K. Sahu, Forester and Shri Resham Lal Verma, Forest Guard, working in Rajnandgaon forest division, the task of making a good beginning in the field of "Joint Forest Management" was accomplished in the forest divisions. The exemplary results exhibited in this field shall provide further impetus to the efforts of Joint Forest Management in future.

Observations

The following observations have been made with respect to these aforesaid committees :—

1. Homogeneous community provides congenial atmosphere for involvement of people in participatory management of natural resources.

2. Resource constraint is the root cause behind the people's awakening towards JFM. People in such sample villages had already an informal village forest committee and they were managing the revenue forests well.
3. People's involvement is high in such sample villages because of higher rate of literacy also.
4. The change in energy consumption pattern indicates the adaptability of resident people towards the changing needs.
5. The role of forest staff in motivating people was also crucial in synergising the community dynamics in such sample villages towards accomplishment of participatory forest management.

Case Study 2 : JFM in Rajnandgaon District: A Short Appraisal

Evaluation of JFM activities on a periodic basis is *sine qua non* for strengthening management practices. Towards this end, an internal system for evaluation has been developed by the forest department to assess the performance of VFC/FPCs in the state. This helps the department to understand the strength and weakness of the present practice. On this basis, measures will be taken to strength the programme further. This is the first such evaluation exercise undertaken in Rajnandgaon forest division. Based on this exercise, a short appraisal has been prepared. Distribution of forests in the study area and JFM profile of Rajnandgaon district have been provided in Tables 6.5 and 6.6.

Table 6.5 : Distribution of forests in Rajnandgaon district

(Area in Hectare)

S. No.	*Forest Uses*	*Area*	*% of total forest area*
1.	Teak Forests	17398.45	7.69
2.	Mixed Forests	186067.177	82.28
4.	Blank & Plantation	18011.578	7.97
5.	Encroachment	4451.54	1.97
6.	Others	200.743	0.09
	Total	226129.488	100

Source : Working Plan of Rajnandgaon Division (Old): Year 1999.

Table 6.6 : JFM Profile in Rajnandgaon District

S.No. Item	*Khairagarh*	*Rajnandgaon*
1. No. of VFC	69	77
2. No. of FPC	46	74
3. Villages covered	115	151
4. Members	16072	21311
5. Male	10159	14412
6. Female	5913	6899
7. Area under VFC/FPC in Hectares	34363.83	27341.27

Source : Forest Department Record, 2003.

An exercise to grade JFM committees of the district was carried out to evaluate the performance of different FPC/VFCs in the year 2003-2004. For grading the committees, five criteria have been adopted. They are: (1) financial resource management, (2) people's participation, (3) forest development, (4) forest protection, and (5) village resource development. Again each criterion has been divided into few objectively verifiable indicators. Performance with respect to each indicator has been evaluated and grading has been carried out. Different weightage has been given to different criteria. There are four grades like excellent, good, average and poor. Evaluation of performance was done by the forest departmental staff by collecting field data. Data were then fed into specially prepared computer software, known as JFM Information System. This information system was designed, developed and operated by FMIS Division, Raipur. Out of all committees in the study area, majority are either average or poor. There are 30 and 96 committees, which have been graded, excellent and good respectively (see Table 6.7).

Table 6.7 : Overall Grading of JFM Committees in Rajnandgaon district

S.No. Grade	*No of committees and % of total committees in ( )*	
	Kheiragarh	*Rajnandgaon*
1. Excellent	–	–
2. Good	1(1)	29 (19)
3. Average	24 (21)	72 (48)
4. Poor	90 (78)	50 (33)
Total	115	151

Source : FMIS Database on JFM Year 2003.

The outcome of the grading exercise has been presented in Tables 6.8, 6.9 and 6.10. Table 6.8 provides scale of grades of committees on different criteria such as financial resource management, people's participation, etc. This shows how many committees from the total 115 committees, are placed in different grades such as excellent, good, average and poor. Tables 6.9 and 6.10 present the number of committees falling in different grades with respect to different indicators within the criterion. Table 6.9 pertains to Kheiragarh division and Table 6.10 pertains to Rajnandgaon division.

Table 6.8 : Performance of committees on different criteria

S.N.	*Criteria*	*No. of committees in different Grades*			
		Excellent	*Good*	*Average*	*Poor*
	Kheiragarh Division				
1.	Financial Resource Management	-	1	57	57
2.	Peoples Participation	-	6	69	40
3.	Role in Forest Development	-	1	54	60
4.	Role in Forest Protection	-	5	54	56
5.	Village Resource Development	-	2	35	78
	Rajnandgaon Division				
1.	Financial Resource Management	-	3	66	82
2.	Peoples Participation	-	46	77	28
3.	Role in Forest Development	-	43	76	32
4.	Role in Forest Protection	-	60	60	31
5.	Village Resource Development	-	5	65	81

Source : FMIS Database on JFM Year, 2003.
Note : EXCELLENT(>80% Marks) ; GOOD (Between 61% to 80% Marks); AVERAGE (Between 40% to 60%) ; POOR (< 40% marks)

Table 6.9 : Grading of Committees in Kheiragarh division : year 2003-04

S.No. Criteria	*Indicators*	*No. of committees under different grade*		
		Good	*Average*	*Poor*
1. Financial Resource Management	Income-Expenses Account	30	52	33
	Progress in Committee Fund Income	6	39	70
	Returning of Loan Amount by Committee Members	1	29	85
	Transparency in Expenses	18	61	36
	Uses of Committee Fund as Loan	2	33	80
2. Peoples Participation	Acceptance of Decision by All members	24	60	31
	Attendance of Women in Meetings	4	56	55

(contd....)

(contd....)

	Organizing Regular meetings	25	57	33
	Role of Women's in Decision making	3	62	50
3. Role in Forest Development	Nationalized Forest Produce	6	65	44
	Non-Nationalized Forest Produce	2	62	51
	Other Works	1	47	67
	Rehabilitation of Degraded Forest and Plantation	15	58	42
4. Role in Forest Protection	Controlling Encroachment	7	54	54
	Controlling Illicit Felling	9	76	30
	Fire Incidences	7	62	46
	Grazing Control	6	47	62
	Nistar and Fuel	10	57	48
5. Village Resource Development	Family employment based Works	0	38	77
	Irrigation Source Development	6	34	75
	Other Works	3	30	82

Note : EXCELLENT(>80% Marks); GOOD (Between 61% to 80% Marks); AVERAGE (Between 40% to 60%) ; POOR (< 40% marks).
Source : FMIS Database on JFM Year, 2003.

Table 6.10 : Grading of Committees in Rajnandgaon division Years 2003-04

S.N.	*Criteria*	*Indicators*	*Grades*		
			Good	*Average*	*Poor*
1.	Financial Resource Management	Income-Expenses Account	8	85	58
		Progress in Committee Fund Income	7	65	79
		Returning of Loan Amount by Committee Members	2	60	89
		Transparency in Expenses	8	82	61
		Uses of Committee Fund as Loan	2	61	88
2.	Peoples Participation	Acceptance of Decision by All Members	52	81	18
		Attendance of Women in Meetings	45	81	25
		Organizing Regular meetings	45	79	27
		Role of Women's in Decision making	45	77	29
3.	Role in Forest Development	Nationalized Forest Produce	47	85	19
		Non-Nationalized Forest Produce	50	83	18
		Other Works	21	89	41

(contd....)

(contd....)

		Rehabilitation of Degraded Forest and Plantation	42	75	34
4.	Role in Forest Protection	Controlling Encroachment	61	60	30
		Controlling Illicit Felling	62	71	18
		Fire Incidences	59	66	26
		Grazing Control	28	86	37
		Nistar and Fuel	58	70	23
5	Village Resource Development	Family Employment Based works	6	71	74
		Irrigation Source Development	5	74	72
		Other Works	4	78	69

Note : EXCELLENT(>80% Marks) ; GOOD (Between 61% to 80% Marks); AVERAGE (Between 40% to 60%) ; POOR (< 40% marks).
Source : FMIS Database on JFM - Year 2003.

Findings of the Case Study

The findings of the case study are presented below :

- VFC/FPCs have started to take necessary steps in forest protection especially against illicit felling, fire protection and uncontrolled grazing.
- Regeneration of forest allotted to JFM committees has improved.
- There is an increase in faith of people on forest department.
- A sense of ownership for forest allotted to the committee is slowly emerging in the mindset of people.
- Attendance of women in the meetings is very poor.
- Village resource development on the basis of 'Shramdan' is non-existent.
- Development works have been done only when financial allotment has been made to the committee.
- Increase in irrigation facilities has received utmost priority in village resource development.
- Majority of VFC/FPCs are not cohesive social organization.

IV

Scaling-up Participation

Scaling up entails replicating into a large-scale implementation process that has the potential to bring desirable outcomes. It also facilitates to make the approach appealing to a wider set of development organizations and improving their capability to adapt them effectively. This section focuses on approaches that the forest department employs. We begin with an examination of the conditions under which forest department adopts participating approaches, and some measures available to overcome the problems. Finally, we emphasize the need to change the way forest department works in order to effectively scale up participating approaches and identify some of the challenges involved. These challenges are: The nature of implementing agencies has strong influence on how participating process should continue. The department makes minimum effort to work with communities. In government department the staff has little time to devote to social organization. They do not generally believe that community participation is needed or desirable. Adequate funds are not devoted to work relating to social organization. Monitoring is also not effective. Most of the supervising staff themselves do not have adequate understanding of participation. The extent of manpower and the skills of staff engaged in social organizations, values held by implementing organizations and the extent and quality of monitoring are some of the critical factors that explain differences in the adoption of participating approaches. These factors reflect differences in capability and incentives of organization to work with communities. Some measures can be taken to facilitate participation processes. Department need to be given additional funds to social organizations. It may be useful to train staff on how to work with communities. Improving monitoring of JFM activities would be useful.

There is a lack of required skills and attitude among the staff. The aspects such as joint decision making, devolving authority, etc. discourage the government staff from facilitating community participation. Working with communities also requires perseverance and humility. This is generally lacking among the staff. Forest staff view themselves as experts with knowledge superior to those of communities, having the rights to prescribe technical solutions. Engaging in joint decision-making grants some of the rights – the rights to information and the right to decide appropriate technical solutions – to communities. It undermines their position of authority and expertise.

The consequences of participation are not also shared equally at all levels and therefore, the decision makers at higher levels in the forest department may be keener than field staff to adopt participating approaches. No amount of exhortations on the virtues of participation can convince the operational staff to overlook the impact of community participation on them. Ensuring that forest departmental staff at ground levels works with communities is, therefore, a challenging task for the managers of the JFM programmes.

Given that supervisors have difficulties in determining whether participation is taking place or not, implementing staff can be expected to make the required effort only under limited conditions. The first is when it is in their interests to do so. The possible motive could be the values they hold which make them favourably disposed to enabling community participation because they feel it is a right way to do their job. The second condition is one when the staff is clearly monitored followed by credible threat of consequences for not adhering to participating approaches. The third condition is one in which the communities demand participation. However, there is a presumption that the communities want to participate. In reality the first two conditions are generally prevalent.

Conclusion

Area-wise coverage of JFM in the state appears to be quite satisfactory in the initial stage. Approximately 2883252 hectares of forest area is currently covered by 6687 JFM committees. This works out to approximately 48 per cent of total forest area under the administrative control of Chhattisgarh Forest Department. In quantitative terms, there is enough scope for expansion of this area too. But what is more important is expansion of the practice in qualitative terms. Success of participatory forest management lies in expanding the practice in the rural milieu and not in expanding the fold of JFM in terms of forest areas. Issues related to gender and equity considerations, empowerment, linkages with Panchayat Raj bodies, conflict management, organisational matter, capacity building of committees members etc. still require greater attention at planning and implementation level. The micro-planning process needs to be made more participatory in order to enhance the level of acceptance of the' micro-plans by the community. Similarly, simplification of various processes and mechanisms to strengthen the village level institutions are also needed. Sharing of information among various JFM committees as well as that between the committees and the forest department is yet another area of concern. The most important aspect is devolution of certain powers regarding forest management and control to the village committees. A beginning has been made, but further dent has not taken place.

7

On-farm Tree Cultivation

Prospects and Challenges

On-farm cultivation of trees is a well-entrenched component of the farming system in Chhattisgarh. Over the years, on account of a dynamic change in rural economy, it has emerged as a private economic resource having immense social security value. Beside this, it has also helped in sustainable development of the existing forest resources in the district Rajnandgaon. The paper presents a strategy to stimulate on-farm tree cultivation in the district.

Farmers in the Chhattisgarh manage trees as an integral part of the farming system. They evolve suitable management practices in a given ecological and socio-economical setting. The objective of such practices is to produce goods like small timber, poles, fuelwood, fodder and other saleable tree products and services like conservation of soil and moisture. More importantly, looking into the cultural milieu of Chhattisgarh, growing of tress in grooves also caters to the religious and cultural needs of the rural folk. As the diversity of goods and services derived from trees become appreciated, a wider potential was postulated for 'agro-forestry'– as a tool for resource poor farmers in stabilising and improving their farm system (Nair, 1984). But what stimulates the rural folk to go for on-farm tree growing is a matter to ponder over. There is no single reason responsible for the emergence of this tree growing culture among the farmers. A dynamic change in the rural economy in the Chhattisgarh region, more especially in the districts of Durg, Rajnandgaon, Raipur, Bilaspur has spurred wide spread on-farm tree growing in the rural landscape.

Study Area and Methodology

The study area occupies the whole district of Rajnandgaon. It is situated in the western part of Chhattisgarh. It extends between 20° 48' and 21° 48' north longitude, 80° 2' and 81° 4' East Longitude. The total geographical area of the district is 9208 square kilometre. Out of it forest spreads in about 2931 square kilometres, i.e. 32 per cent of the total geographical area. The gross cropped area is 4722.66 square kilometres. The balance area has been categorised as wasteland and non-agricultural land (Table 7.1). The district has a population of 10.89 lakh. The density of the population in the district works out to 128 per square kilometre. The district is rich in physical and forest resources. It receives an average rainfall of about 1300 mm from the south-eastern monsoon. The district is rich in agricultural resources too. It falls in the rich rice belt of Chhattisgarh.

Table 7.1 : Land use Classification of District Rajnandgaon

S. No.	*Item*	*Area in square km*
1.	Total Geographical Area	9207.57
2.	Forest Area	2931.02
3.	Area not available for cultivation (A+B)	2024.86
	A. Area under non-agricultural use	1948.95
	B. Barren and uncultivable area	75.91
4.	Fellow Land excluding Waste land	139.85
5.	Waste land	460.5
6.	Net Area sown	3651.34
7.	Gross Cropped Area	4722.66
8.	Area sown more than once	1071.32

Source : District Statistical Book, 1997, Rajnandgaon.

The distribution of land holding presents a distinct picture of distribution of agricultural land among the rural folk. It is observed that 38.3 per cent of all operational holding falls in the size less than one hectare covering about 7.5 per cent of the total area under holdings. About 22.4 per cent holdings are in the size class of 1 to 2 hectares which comes 13.7 per cent of the area. About 21.5 per cent of holdings are in the size class of 2 to 4 hectare which comes 21.4 per cent of the area. The largest holding size of 4 to 10 and above is comprised 18.7 per cent of the holdings and cover about 61.06 per cent of their area (Table 7.2).

Table 7.2 : Distribution of operational holdings and area operated in 1996-97

S.No.	*Category of holdings*	*No. of operational holdings*	*Percentage of Area operated*
1.	Marginal [Below 1 half]	13092 (38.3)	7.5
2.	Small [1 Hect. to 2 Hect.]	7635 (38.3)	13.7
3.	Semi-medium [2 to 4 Hect]	7337 (21.5)	21.4
4.	Medium [4 to 10 Hect.]	5056 (14.8)	40.86
5.	Large [10 Hect. and above]	.998 (3.9)	20.2
	TOTAL	34118	

Note : Figure in bracket indicates the percentage.
Source : District statistical book Rajnandgaon year 1997.

Cultural Ecology and On-farm Tree Management

Historically Chhattisgarh was a land of forests. Farm lands were curved out of the forests. Over the years forest trees were gradually cleared to pave way for agriculture. With the passage of time suitable multipurpose indigenous trees appeared on the farm as a sequel to growing fuel wood and fodder needs of the region. However forest trees are still found scattered on the recently reclaimed farms situated in proximity to forests. Apart from this, a lot of land was used as fellow in the past. With the increase in pressure on land, more and more land has been brought under agricultural practice. However on such land also either existing forest trees are retained or indigenous species are cultivated. Therefore, two distinct category of areas have emerged in the district. The category-1 represents area which are more than 10 km away from the existing forests. The category-2 area represents those areas which are in proximity to the forests, i.e. with in 10 km radius of the forests. Density of on-farm trees is more in category-1 than that of category-2 area. Apart from this, category-1 area contains trees grown by the farmers; where as in category-2 area which have been reclaimed from the forests in the post-Independence era, standing forest trees are retained and managed by the farmers. Therefore, retention and cultivation of trees on farm land far off the forests as well as along the forests appear as a strong practice evolved as a response to a dynamic change in the rural sector.

The three main species obsessively grown by the farmers in this tract are Babool [*Acacia arabica*], Siris [*Albizzia procera*], and Arjun [*Terminalia arjuna*]. The first two tree species are leguminous plants and have light crown. All the three are multiple-use trees and so they provide various tree products like fuelwood, small timber and fodder. Having trees *in situ* on the farms, also provides some of their concomitant intangible benefits previously provided by the surrounding forests. Such benefits are in the shape of shade from the scorching sun, protection against torrential rain or soil loss on account of surface run-off.

Beside the factor of intensification of agriculture there are other factors which have contributed to the phenomenon of increasing tree density on the farm land. Apart from this, increasing population pressure, socio-economic changes, greater marketisation of products have substantially contributed to increased differentiation and fragmentation within rural communities which ultimately led to unregulated use of village forests. As a result village forests vanished from the rural scene in most of the district of Chhattisgarh. Gradual reduction in the availability of common land tree resources in the rural areas created a situation which compelled the farmers to grow trees on their farm. Initially on-farm tree growing was a response to increasing fuel wood crisis. However, over the years a dynamic change in the socio-economic scenario in the rural sector coupled with fuelwood crisis stimulated development of on-farm tree resources.

Shift to Private Tree Resources

With the forest resources depleting fast, population increasing alarmingly, village common land tree resources completely decimated and prices of timber and other tree products increasing leaps and bound, there is a marked shift towards private tree resources. Farmers have shown enormous interest in growing and cultivating selected trees of value to them on their land. The expanding market for fuelwood and other tree products in the region has accelerated the process of on-farm tree growing. It is because of the fact that tree apart from being a potential source of small timber and fuel wood, also emerged as a cash crop to farmers.

The system of intensive crop agriculture on account of abundant rainfall and sunshine in this tract commonly encourages retention and growing of tress not only within the fields but also along the boundaries and other niches not used for crop. However, if one cruises along the North-South plane of the district different patterns of tree growing become conspicuous. The northern part which relatively receives less rainfall has less tree density than its southern counterpart. Another pattern of spatial distribution of tress on the farms in relation to its proximity to forests emerges. Farmlands situated near the forests, i.e. within the radius of five to ten kms from the forest boundary possess conspicuously less tree density than the areas far away from the forests. More to it, on the farms nearer to forests wild trees once a part of the forest ecosystem stand conspicuously. In other words, what is noteworthy is the fact that in those areas where forest areas are plentily available, on-farm tree growing and management is either non-existent or distinctively passive. However, intra-district agro-ecological variation notwithstanding, on-farm tree cultivation practices in the area away from the existing forest tract emerges strongly in the rural landscape of the district. In such areas common tree management strategies such as retention of naturally obtained tree seedlings, pollarding, debranching, pruning, coppicing, etc. are widely adopted.

Dynamics of Rural Change

The rural sector is undergoing a social and economic change. On-farm tree cultivation is a reflection of economic response on the part of farmers to this change. Therefore in order to understand the phenomenon of on-farm tree cultivation in its proper perspective, it is imperative to study the dynamics of rural change.

What is of paramount significance in the economic life of the farmers in relation to various tree products is their response to changes in demand and supply and to scarcity and abundance. Solution to problems of scarcity of tree products in areas away from the forests have been evolved in the form of either on-farm tree growing practices or decreased dependence on forest biomass or with adopting other available alternative. A study was conducted in three villages of the district. These villages were Maldongri, Maregaon, and Sangli of Chowki Block. All the study villages in the span of 30 years period have shown increased adaptation to other sources of energy like use of cowdung cake, switching over to use of kerosene stoves and biogas plants. Beside this, the villagers among the study villages have shown attitudinal changes towards energy efficiency by having gradually adopted aluminium utensils in place of earthen pots for the purpose of cooking.

Therefore, looking at the pace of socio-economic transformation in the rural areas of the district it is reasonable to assume that in the future more such changes will take place. With improved access to market, on account of social, political, economic and technological change in the rural sector, outlets for livestock products like milk and ghee are fast created. If this gains further momentum then it will be in the realm of possibility in the near future that livestock management may shift towards pasture crops and stall feeding rather than allowing cattle to graze in the beleaguered village common land and in the nearby forests. As the market opportunities for tree products are improving, the time is not far away when there will be a widespread emergence of alternatives which lead to lowering of opportunity costs to the farmers in trading the total trees of a Babool or Siris or Arjun. Such alternatives have partially emerged in the district after the State Government's order relaxing Transit Pass Rules for such tree species. As things stand today, preferences for such alternative have emerged among the farmers. In few intensively studied villages, it was also found that up to sixties there was an unhindered supply of fuel wood from the nearby forests. But afterwards the nearby forests were completely degraded. This has set off a change in nature of fuel used by the households in the villages. The farmers have started using cowdung, crop residues, stalks of Ipomea and wood of Palas tree. What is significant is the change in the perception of the farmers towards usability of species like Ipomea and Palas tree having low calorific value. Palas is a on-farm tree which spreads in the farm naturally by its root suckers. Earlier this tree was not used as fuel on account of its low caloric content. But with the decreased availability in fuelwood from the forests, and increased marketisation of tree products, the tree Palas (*Butea monosperma*), growing naturally on the crop field is being used as fuel, making it possible on the part of farmers to maximise tradable on-farm tree products. It is generally seen that farmers enter the market when returns from tree crops appear to be assuring, attracting and stable. Market formation with respect to tree products especially small timber has substantially improved consequent upon the liberalisation of forest and transit rules in the state. This is amply evident from the fact that 80 per cent of the saw mills in the district were revived on account of uninterrupted supply of wood from the farms in the period 1991-97. Otherwise these saw mills were closed on account of paucity of timber. As things stand today, the supply of timber to the private saw mills from farm forests is more than that of Government Depot (Table 7.3).

Table 7.3 : Quantity of Timber handled in saw mills in Rajnandgaon District. Period : 1992-97

Sl. No.	*Source of timber*	*Quantity of timber in cmt*
1	Government Depot	38890
2	On Farm trees	63745

Source : M.P. Forest Department Record 1997.

The second most important economic change visible in the village life is their perception towards resources. Farmers in the category-1 area have accepted on-farm trees as an economic resource. A more holistic and dynamic framework for understanding such change is provided by the concept that households use the resources available to them to pursue a strategy that includes food security, social security and risk management, as well as income generation, among their objectives (Chambers, 1983). Overtime, both objectives and resource availability will change, necessitating adaptations in the strategy the household pursues. The choices farmers make between different courses of action will also be influenced by changes in socio-economic conditions overtime that alter the cost and returns they face (Ruthenberg, 1980). What is of immense significance is the fact that tree growing has been accepted by the farmers as livelihood security measures. In one way it is also an induced innovation in response to the dynamic change in the rural sector in the recent times. The objective of responses evolved in the rural landscape are as follows :

1. To maintain supplies of tree products as production from nearby forest tree resources has declined due to deforestation.
2. To meet growing demands for tree products as population grows, new uses for tree outputs emerge or urban markets develop.
3. To contribute to risk reduction and risk management to tide over breaks in seasonal flow of produce and income and to provide a reserve of tree biomass resources as a capital for use in the time of stress, contingencies and emergencies.

Evolving a Strategy to Stimulate On-farm Tree Growing

The option of on-farm tree growing in category-1 area, i.e. those areas which are away from forests has taken deep roots in the psyche of the farmers. This became possible because of the fact the farmers respond to the present situation in favour of tree growing on the farms. However, the tree growing option is subjected to change in the relative prices and availability of farmers' resources of land, labour, capital and other inputs and outputs. Today trees are standing on the cropland because of the fact that more than one of the reasons elucidated above has made it possible. A combination of factors which influence farmer's decision about tree growing and management always works in tandem. These factors are the availability, cost and allocation of land, labour and capital; the impact of subsistence demand and market opportunities; and farmer's attitudes to risk and management of different form of risk (Arnold 1997). A perusal of land holding pattern in the district underscores the fact that the medium category of land holding constitutes the majority of farmers. This holds immense possibility for on-farm tree cultivation.

Nevertheless, over the years, expanding market opportunities for tree products have played significant role in influencing the farmer's decision regarding on-farm tree growing practices. However, this practice is widely witnessed in those category-1 areas. In category-2 areas, i.e. areas which are situated near forests, tree growing has not emerged as an economic option. It is because of the fact that farmers in the category-2 areas are still accessible to nearby forests to meet their needs. Apart from this, forestry operations still provide partial livelihood security, cover and cushion for risk reduction.

Over the years, the strategy to stimulate on-farm tree growing did not get the appropriate focus. Rather, it became more influenced by phenomenon and concerns like deforestation, increasing population, unemployment, poverty, environmental degradation and meeting the fuel wood, small timber needs of rural folk. Each of these perspectives provides some intrinsic insights into one or more possible roles tree play in the rural household livelihood. But, if different aspects of tree growing are analysed and pursued in isolation to each other, without considering the broader socio-economic framework within which trees are positioned, the strategy to stimulate on-farm tree growing is bound to fail. Examples of failures are galore. The scheme of "Brikshya-Sahakarita" a social forestry programme which aimed at increasing the livelihood security of the marginal and landless farmers in the rural sectors has failed to take off. In the past many tree planting support programme have also failed on account of poor technical prescription and extension services and inadequate understanding of role of institutions. Implementation of such schemes was forced to run ahead of capability to provide adequate extension and technical packages. Apart from this, there was no effort on the part of government to link production to market. In the previous scheme known as "Decentralised Nursery" project intervention mostly centred around provision of subsidised planting stock and/or cash payment to localised nursery holders to offset establishment and maintenance cost.

Subsidised supply of fuel wood and timber in the areas either near or away from forests dampens enthusiasm for on-farm tree growing. As things stand today, forest produce having a value of 26 million are being distributed among the villagers. Out of it, the subsidy component amounts to 18.5 million (Table 7.4). Apart from this the earlier programme of subsidising at the level of initiating the tree growing movement by distributing free seedlings also did not work , because production has never been linked with stable markets. Therefore, a subsidy at the level of distributing seedling in both categories-1&2 areas shall work if it is linked to emergence and formation of a stable market. Market formation and development is again influenced by host of factors, out of which few lies within the ambit of government and rest lies in the realm of economy. However, over the last decade, decreased supply of forest products from the natural forests, increase in size of middle class in the society, industrialisation and simplification of forest transit rule and land revenue laws have helped in developing and shaping tree product market in the state.

Table 7. 4 : Forest produce distributed to people of *Rajnandgaon district* by forest department under Nistar and consumer service scheme: Year 1995-96

S.No.	*Name of forest produce*	*Quantity*	*Market value in lakh*	*Value at which supplied/distributed in lakh*	*Amount of subsidy in lakh*
1.	Fuelwood	296.05 Ton	26.464	12.13	4.334
2.	Timber	136.393 cmt	7.69	6.41	1.28
3.	Poles	26530 No.	12.81	6.796	6.014
4.	Bamboo	2302778 No.	222.215	48.52	173.695
	Total		259.179	73.856	185.323

Source : M.P. Forest Department Record, 1996.

Another way of stimulating on-farm tree growing practices will be to provide credit for nursery development to "Village Forest Committee and Forest Protection Committee" which have been formed in villages in category-2 areas with a broad aim of protecting and developing the forest under the broad policy framework of participatory management of forests. After having started micro forest nursery development programme basing on participatory model, the next steps, which should be addressed, are on the market and demand side. Therefore, more attention is required to be paid to policies, practices and procedures that currently limit the access to market and depress market prices for different tree products. These commonly include lack of market formation, poor functioning trading systems serving small farmers, competition from supplies from government forests and restriction on harvesting of wood products on private land. Putting it in other way, government, by taking action or by not making proper intervention, not only creates an environment where such factors acting in tandem hinder farmer's access to tree product markets but also inadvertently interferes with the process of change that is driving the rural economy of this belt from a subsistence economy to market economy.

Therefore, the basic elements of the strategy to stimulate on-farm tree growing practice in the district Rajnandgaon should comprise the following elements :

1. Identifying and delineating the two categories of areas, i.e. Categories-1 & 2 area and enumerating the villages falling in both the category areas (Fig 7.1).
2. Extending credit facilities to VFC and FPC in Category-2 areas to develop nurseries.
3. Linking subsidy in planting or growing on-farm trees to Category-2 areas only. The concept of cost sharing can be linked to subsidisation programme.
4. Developing market support establishing few nuclear timber and fuel depot. in both Categories-1 & 2 areas. These depot should function as procurement as well sale depot. In addition to it, there should not be any restriction on development of private markets. However, the role of middle men has to be totally eliminated to make on-farm tree attractive and economically feasible. The only objective of market intervention of government should be to create congenial

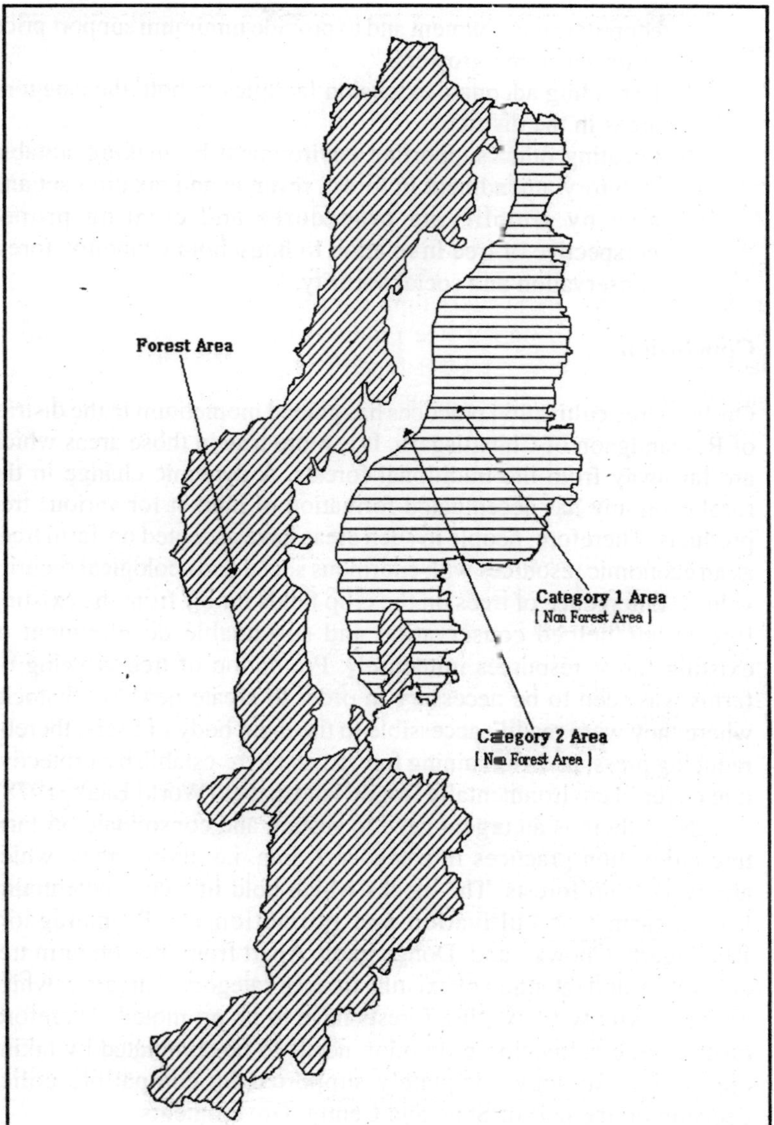

Fig. 7.1 : Map Showing Forest Area and categories I and II of Rajnandgaon district

economic environment and to provide minimum support price to on-farm tree growers.

5. Extending adequate extension facilities to both the category areas in the district.
6. Creating other supporting environment by making suitable statutory amendment in forest, revenue and taxation act and also by simplifying procedures and creating proper perspective of tree in relation to household economy, forest conservation and social security.

Conclusion

On-farm tree cultivation practices has gained momentum in the district of Rajnandgaon of Chhattisgarh. It is more true in those areas which are far away from the traditional forests. A dynamic change in the rural economy has accentuated formation of market for various tree products. Therefore, people in such areas have accepted on-farm trees as an economic resources with enormous social and ecological security values. Emergence of trees on the crop fields far off from the existing forests has helped conservation and sustainable development of existing forest resources immensely. Promotion of tree growing on farms was seen to be necessary in order to create new wood stocks where they were readily accessible to the main body of users, thereby reducing pressure on remaining forests, and to re-establish a protective tree cover in environmentally fragile landscapes (World Bank, 1978).

Now there is an urgent need to sustain and consolidate on-farm tree cultivation practices in category-1 area, i.e. those areas, which are away from forests. The blocks which hold immense potentiality for on-farm tree cultivation and promotion are Rajnandgaon, Khairagarh, Chowki, and Dongargaon. Apart from this, on farm tree cultivation and retention of existing trees in category-2 area, i.e. which are in proximity to existing forests have to be promoted. Therefore, on-farm tree cultivation promotion needs to be stimulated by taking appropriate strategy adequately supported by compatible policy decision on the part of State and Central Governments.

APPENDIX I

List of Important Tree Species found in Rajnandgaon District

Sl. No.	*Common Name*	*Botanical Name*
1	Amla	*Emblica officinalis*
2	Babul	*Acacia nilotica*
3	Bahera	*Terminalia belerica*
4	Bel	*Aegle marmelos*
5	Bhelwa	*Semecarpus anacardium*
6	Bhirra	*Chloroxylon swietenia*
7	Bhoursal	*Hymenodictyon excelsum*
8	Bija	*Pterocarpus marsupium*
9	Dhawda	*Anogeisus latifolia*
10	Dhobin	*Delbergia paniculata*
11	Garahi	*Cleistanthus collinus*
12	Haldu	*Adina cordifolia*
13	Harra	*Terminalia chebula*
14	Jamun	*Syzygium cumini*
15	Kala Siris	*Albizia lebbek*
16	Karanj	*Pongamia pinnata*
17	Karra / Landia	*Lagerstroemia parviflora*
18	Kullu	*Sterculia urenus*
19	Kusum	*Schleichera oleosa*
20	Lendia	*Lagerstromia parviflora*
21	Mahua	*Madhuca indica*
22	Neem	*Azadiracta indica*
23	Nilgiri	*Eucalyptus species*
24	Palas Bel	*Butea superba*
25	Saja	*Terminalia tomentosa*
26	Sal	*Shorea robusta*
27	Salai	*Boswellia serrata*
28	Saphed Siris	*Albizia procera*
29	Shisham	*Delbergia latifolia*
30	Teak	*Tectona grandis*
31	Tendu	*Diospysos melanoxylon*
32	Tnsa	*Ougeina oogeinensis*

APPENDIX II

Government of Chhattisgarh Forest and Culture Department Mantralaya, Raipur

No. F 7-42/2001/F.C. **Raipur, the 22nd October, 2001**

RESOLUTION

Subject : Chhattisgarh State Forest Policy, 2001

1. Preamble

1.1 The GOI *vide* resolution No. 3-1/1986/F.P. dated the 7th December, 1988, enunciated a National Forest Policy that provided a national perspective on forest management problems and specified actions required for tackling them.

1.2 The new state of Chhattisgarh came into being on 1st November, 2000. There is a necessity to review the situation and to evolve for the future, a new strategy of forest conservation, which encompasses the special characteristics of the state as well as provides it a new direction. Since forest is a concurrent subject under the Constitution of the country, the state's forest policy has to remain in tune with the current National Forest Policy.

1.3 The state of Chhattisgarh, lying between 17° 46' N to 24° 6' N latitude and 80° 15' E to 84° 51' E longitude, has about 44 per cent of its geographical area (135,224 sq. kms) under forests and provides catchment to at least four main river systems, i.e. Mahanadi, Godavari, Narmada and Ganges. Major rivers of the state are

Mahanadi, Indravati, Hasdeo, Sheonath, Arpa and Ibb. The climate of the state is generally sub-humid with an annual rainfall ranging from 1200 to 1500 mm.

1.4 Over the years, the forests in the state have suffered serious depletion. This can be attributed to relentless pressures arising from ever-increasing demand for fuelwood, fodder and timber; inadequacy of protection measures; diversion of forest lands to non-forest uses without ensuring compensatory afforestation and essential environmental safeguards; and the tendency to look upon forests as a revenue earning resource.

1.5 The forests of the state fall under two major forest types, i.e. Tropical Moist deciduous forest and the Tropical Dry deciduous forest. Sal (*Shorea robusta*) and Teak (*Tectona grandis*) are the two major tree species in the state. Other notable overwood species are Bija (*Pterocarpus marsupium*), Saja (*Terminalia tomentosa*), Dhawra (*Anogeissus latifolia*), Mahua (*Madhuca indica*), Tendu (*Diospyros melanoxylon*) etc. Amla (*Embilica officinalis*), Karra (*Cleistanthus collinus*) and bamboo (*Dendrocalamus strictus*) constitute a significant chunk of middle canopy of the state's forests.

1.6 Biogeographically, the state falls in Deccan *bio*-region comprising representative fauna of central India like the tiger (*Panthera tigris*), leopard (*Panthera pardus*), gaur (*Bos gaurus*), sambhar (*Cervus unicolor*), chital (*Axis axis*), nilgai (*Boselaphus tragocamelus*) and wild boar (*Sus scrofa*). The state is a proud possessor of rare wildlife like the wild buffalo (*Bubalus bubalis*) and hill myna (*Gracula religiosa*) which have been declared as the state animal and bird respectively.

1.7 The state is richly endowed with mineral resources like the coal, iron, bauxite, limestone, corundum, diamond, gold, tin, etc. which fall mainly within the forests of the state.

1.8 The population of the state is predominantly tribal, who have significant economic and cultural dependence on the forests of the state. There is a large population of non-tribal landless and economically backward communities in the state who derive livelihood security from the forests of the state.

2. Basic Objectives

2.1 The basic objectives that should govern the State Forest Policy are the following:

- Unlocking of the vast array of forest resources on sustainable basis for enhanced well-being of local people by converting these open access resources (OAR) into community controlled, prioritized, protected and managed resources.
- A shift in accent from major to minor forest produces, from crown to multi tier forestry and from flagship species to smaller denizens of the forests.
- Maintenance of environmental stability through preservation and where necessary, restoration of ecological balance that has been adversely disturbed by serious depletion of forests in the state.
- Conserving the Bio-cultural heritage of the state by preserving the biologically rich natural forests that provide the essential cultural milieu to the tribals of the state.
- Checking the denudation of forests and soil erosion in the catchment area of the rivers, and reservoirs for soil and water conservation; mitigating the floods and droughts; recharging of water bodies, aquifers and for the retardation of siltation of the reservoirs.
- Increasing the forest/tree cover in forest deficient districts through afforestation and agro forestry/farm forestry programmes, especially on all denuded, degraded and unproductive lands.
- Meeting the requirements of fuelwood, fodder, minor forest produce and small timber of the rural and tribal population with due regard to the carrying capacity of the forests.
- The derivation of direct economic benefit from the forests of the state shall be subordinated to the requirements of the environmental stability and maintenance of ecological balance in the state.
- Creating appropriate policy and legal framework for the achievement of these objectives.

3. Essentials of Forest Management

3.1 Existing forests and forest lands should be fully protected and their productivity increased. It is necessary to promote efficient methods of timber harvest and utilisation to maximize economic returns from the forests.

3.2 The network of national parks, sanctuaries, biosphere reserves and other protected areas should be strengthened and extended adequately for the conservation of total bio cultural diversity in the state.

3.3 Targeting on broad range of goods and services in terms of physical, material, human social, cultural and environmental assets in conjunction with appropriate entitlement regime, People's Protected Area (PPA) envisions a proactive and people's friendly framework to ensure long term protection and maintenance of biological diversity and providing at the same time a sustainable flow of natural products and services to meet local community needs. Therefore, a network of PPAs should be established as poor people's pool of assets for strengthening livelihood security of forest dwellers.

3.4 Provision of sufficient fodder, fuel and small timber to local people, especially in areas adjoining forests, is necessary to prevent further depletion of forests beyond their sustainable capacity. As fuel wood continues to be the predominant source of domestic energy in rural areas, the programme of afforestation should be intensified with special emphasis on augmenting fuel wood production to meet the requirements of the people. Furthermore, to reduce the pressure on forests due to increasing demand for fuelwood, its substitution by alternative sources of energy should be promoted.

3.5 Minor Forest Produce (MFP) including medicinal plants provide sustenance to the tribal population and to other communities residing in and around the forests. Such produce should be protected, improved and their non-destructive harvesting methods enhanced with due regard to providing employment and income generation opportunities to the dependent people. MFP is the major source of livelihood of tribals and other forest based rural communities. Therefore, rather than exporting MFP in raw form, efforts should be made, as far as possible, to promote processing and value addition of the same, at the local level.

3.6 Supply of timber and poles to urban centers from non-forest sources is necessary to reduce pressure on natural forests. Therefore, state government should encourage agro-forestry, farm forestry and on-farm cultivation of timber trees.

3.7 Establishment of appropriate instruments including policy and legislative measures to protect the rich bio-cultural heritage of the state in view of increasing threats of bio-piracy and infringement of IPR (Intellectual Property Rights) from within and outside the state.

3.8 All forest areas in the state should be managed in accordance with a duly approved management/working plan.

4. Strategy

The objectives and essentials of forest management for the state should be achieved through a well-defined strategy as under:

4.1 Area Under Forests

The national goal is to have a minimum of one-third of the total land area of the country under forest or tree cover. Though the state has the rare distinction of having more than one-third of its geographical area under forest, there are few districts, where the forest area is less than the norm, and where there is a need to expand the forest cover in addition to preserving the existing forest cover.

There is a necessity to preserve the existing tree cover on hill slopes and landscapes providing catchment to rivers originating or flowing through the state.

4.2 Management of State Forests

4.2.1 No forest should be permitted to be worked without a

duly approved working/management plan, which should be in a prescribed format and in keeping with the National Forest Policy/State Forest Policy and the principles of sustainable forest management. The effects of forest management on forests should be periodically measured with the help of set criteria and indictors (C&I). The state should issue necessary guidelines to put in place a monitoring mechanism to ensure regular compliance of management/working plan prescriptions.

4.2.2 In order to meet the growing needs of people for the essential goods and services that the forests provide, it is necessary to enhance the forest cover in forest deficient districts and to increase the productivity of the existing forests through appropriate scientific and technical inputs.

4.2.3 No exotic species should be introduced, through public or private sources, unless long-term scientific trials undertaken by specialists in ecology, forestry, sociology and agriculture have established that they are suitable and have no adverse impact on indigenous vegetation, ecology and bio-cultural environment of the state.

4.2.4 Joint Forest Management (JFM) practices should form the basis of forest management in the state. Necessary provisions should be made for the adequate participation at all levels of decision making by the landless, marginal farmers and women in all JFM bodies like the VFC (Village Forest Committee), FPC (Forest Protection Committee) and EDC (Eco Development Committee). The abundant potential of people living in rural and forest areas should be tapped for sound participatory forest management. Efforts should be made to facilitate assistance from financial institutions to the forest dwellers engaged in forest based economic activity for furthering participatory forest management in the state.

4.2.5 For sustainable forest development, livelihood security and bio-cultural diversity conservation, People's Protected Areas (PPAs) should be established. This paradigm shift of adaptive management can reconcile the dichotomy of threat perception arising out of conservation-development orthodoxy by taking into account the human sensitivities like their socio-cultural norms, beliefs and systems borne out of history, culture and traditions.

4.3 Rights and Concessions

Forests of the state recognize traditional rights and concessions of entry into forests and use of the produce thereof, by the people living in and around the forest areas. Such rights and concessions, popularly called 'Nistar', have sometimes legal as well as customary basis and the state is obliged to provide for them. In due course of time such rights and concessions with the exception of cultural rights, may no longer be required with an improvement in the standard of living of the majority of people in the state.

4.3.1 The provision of Nistar, should always remain related to the carrying capacity of the forests. The capacity itself should be optimized by increased investment in silvicultural research and socio-economic development of the area. Stall feeding of cattle should be encouraged. The requirements of the community, which cannot be met from the existing forests, should be met from plantations under social forestry/agro forestry/farm forestry in areas outside the natural forests.

4.3.2 The holders of customary rights and concessions in forest areas should be motivated to identify themselves with the protection and development of forests from which they derive these benefits. The rights and concessions from forests should primarily be for the *bona fide* use of the communities living within a radius of five kilometres from the existing natural forests. It is visualized that the JFM practices will motivate people to keep their customary rights and concessions on forests within the carrying capacity of the forests.

4.3.3 The socio-economic and cultural life of tribals and other communities living within and near forests revolves around the forests. Their domestic requirements of fuelwood, fodder, minor forest produce and construction timber should be the first charge on the forests of the area. The rights and concessions enjoyed by them should be protected, with due regard to the demands of the conservation of biological diversity in the area.

4.4 Management of Sal and Bamboo Forests

Sal and bamboo forests in the state constitute an important component of the forest ecosystem of the state. The state has large chunks of ecotone forests between Sal and miscellaneous forests requiring special management practices. Such forests are not only ecologically sensitive, but also provide Bamboo and other basic needs that constitute the essential elements of the livelihood security of the poor and tribal people of the state. Therefore, special treatment of ecotone Sal forests and the restoration of the degraded Bamboo forests as well as the maintenance of good Bamboo forests should be the state's priority.

4.5 Conservation of Minor Forest Products (MFP)

Non-timber forest produce called the Minor Forest Products or MFP like Tendu leaves, Sal seed, Imli, Chironji, Kullu and Dhawra gum, Kosa cocoon, Honey etc., form an essential element of the means of livelihood of the tribals and the landless, marginal farmers and other rural poor communities of the state. It is often suggested that it is the MFP and not the so called major forest produce like timber that is the mainstay of the rural poor. MFP like the Tendu *patta* (leaf) and Sal seed also add sizeable revenue to the state exchequer, which is now distributed among the gatherers.

4.5.1 The state should take appropriate measures through the Chhattisgarh State MFP (Trade and Development) Cooperative Federation Ltd. for sustainable utilization and long term conservation of all MFP found within the forests of the state.

4.5.2 The state should take necessary steps for endowing the ownership rights of MFP on local communities as per the provisions of the Panchayat Upbandh (Anusuchit Chetron ka Vistaar) Adhiniyam, 1996.

4.6 Conservation of Medicinal Plants

Forests have been the source of invaluable medicinal plants since the time man realized their preventive and curative properties and started using them for human health cover. In view of the richness of medicinal and herbal plants in the state, a mechanism should be developed for *in situ* and *ex situ* conservation, domestication and non-destructive harvesting with the active support from local people including traditional healers and *vaidyas*. The socio-cultural, spiritual and medicinal arena of the rural populace particularly the tribal should form the backbone of community based conservation and utilisation of medicinal and herbal plants.

4.7 Protection of Forests

Forests being an Open Access Resource (OAR) are vulnerable to various kinds of pressures like theft, fire, illegal grazing and encroachment. Theft of forest resources like timber, and animal parts and products is considered as high return and low risk offence. Uncontrolled forest fire result in significant loss of biodiversity, loss of forest regeneration, burning of biomass and destruction of micro organisms necessary for essential forest ecological processes. Unregulated grazing by livestock inside forests is a major cause of forest degradation and decrease in its regeneration. Encroachment on forest lands for agricultural diversion has been a major cause of forest loss. Therefore, there is a need to strengthen forest protection measures.

4.7.1 Protection mechanisms should be strengthened by involving local people through village level committees. These committees should be empowered and provided special incentives to prevent the forest offences.
A 'Forest Crime Bureau' with an adequate legal and statistical base should be established for a systematic tackling of crimes and criminals in the forest areas.

4.7.2 Steps should be taken for the establishment of special courts at the district level for quick disposal of forest offence cases.

4.7.3 Grazing regulation in forests should be made more effective through community participation. Grazing by livestock inside the forests should not exceed the carrying capacity of the forest.

4.7.4 Fire in forests should be strictly controlled. Improved and modern techniques for forest fire prevention and control as well as tools like GIS (Geographical Information System) and remote sensing should be utilized for fire control.

4.8 Diversion of Forest Lands for Non-forest Purposes

4.8.1 Forest land or land with tree cover should not be treated merely as a resource readily available to be utilized for various projects and programmes, but as a state asset which requires to be properly safeguarded for providing sustained benefits to the entire community. Diversion of forest land for any non-forest purpose should be subjected to the most careful examination by specialists from the standpoint of ecological, environmental and social costs and benefits. Projects, which involve diversion, should provide in their investment budgets, funds not only for prior regeneration/compensatory afforestation, but also for the development of social infrastructure in the area.

4.8.2 Beneficiaries who are allowed mining and quarrying in forest land and in land covered by trees should be required to adopt cluster approach in mining and to repair and re-vegetate the area, after having used it, in accordance with the Government of India guidelines and established forestry practices. Rehabilitation of the mined areas should be done to ensure ecological restoration of the affected site.

4.9 Bio-cultural Diversity Conservation

4.9.1 The State is extremely rich in its bio cultural diversity. This diversity should be preserved through action as under:

- Intensification of surveys and inventorization of bio-cultural resources in different parts of the state. The survey should include information on the distribution pattern of various species/population/communities and the status of ethnobiologically important groups.
- Conservation of biodiversity through the establishment of a representative network of protected areas including Biosphere Reserves, National Parks, Sanctuaries, Gene

conservation centres, and People's Protected Area. Such areas should cover sites of exceptional taxonomic and ecological value in terms of flora and fauna with adequate emphasis on the lower vertebrate, invertebrate and micro flora, which are important for the maintenance of healthy ecosystems. Tribals and the rural people displaced, if any, due to creation of such national parks/biosphere reserves or gene conservation centres should be fully and properly rehabilitated on such sites and in such manner that their standard of living after the rehabilitation is markedly improved.

- Legal and administrative measures should be taken for the protection of state's bio-cultural diversity against bio piracy and for sustainable use of plant and animal genetic resources. Intellectual Property Rights (IPR) of the people of the state specially the tribals should be zealously guarded. Domesticated species/varieties of plants and animals should be conserved as an integral part of the state's rich genetic diversity.
- Crucial corridors between national parks, sanctuaries, forests and other protected areas should be identified and notified for linking them to maintain genetic continuity of flora and fauna. Such areas should be managed with prescriptions favouring wildlife requirements like the retention of snags, natural gaps, grassy areas, special lithic habitats, caves, cliffs, den sites and water bodies etc.
- Modern techniques of *ex situ* conservation like 'tissue culture' and biotechnology should be promoted for the preservation of endangered and threatened species of wild flora and fauna.
- Monoculture and planting of exotic floral species should be avoided unless sufficient experimentation on strict scientific lines has established their usefulness. Exotic faunal species should not be introduced into the forests of the state.
- Tribals and other indigenous people of the state, residing in and around forest areas, with rich cultural traditions and practices, should be encouraged to maintain their unique relationship with the forests for mutual benefit. Unique geographical and cultural landscapes existing in protected areas should be managed keeping in view the conservation of bio-cultural diversity of the state.

4.9.2 Forest management should take special care of the needs of bio-cultural diversity conservation of the state and the forest management/working plans should include specific prescriptions for this purpose. Wildlife management plans should be prepared for each protected area of the state. Biotic pressures on protected areas should be managed through eco-developmental activities with the active involvement of the local communities.

4.10 Afforestation, Social Forestry and Farm Forestry

4.10.1 A need-based and timebound programme of afforestation and tree planting, with particular emphasis on fuelwood and fodder development, in all the forest deficient districts of the state is an urgent necessity to meet the growing needs of the forest dependent sections of the society specially the landless and those identified as BPL (Below Poverty Line).

4.10.2 Planting of trees along side of roads, railway lines, rivers, streams and canals, and on other unutilized lands under the state, corporate, institutional or private ownership should be encouraged. Green belts should be raised in urban/industrial /mined out areas. Such a programme will also help to improve the microclimate of the concerned area.

4.10.3 The village and community lands not required for other productive uses, should be taken up for the development of tree crops and fodder resources. Technical assistance and other inputs necessary for initiating such programmes should be provided by the State Government, public sector undertakings and agricultural universities.

The revenue generated through such programmes should belong to the panchayat where the lands are vested in them. In all other cases, such revenue should be shared with the local communities. The vesting, in individuals, particularly from the weaker sections (such as landless labour, small and marginal farmers, scheduled castes, tribals, women and BPL) of ownership rights over trees, should be considered, subject to appropriate regulations. Beneficiaries should be entitled to usufruct from the trees and in turn be responsible for their safety and maintenance.

4.10.4 The State Land Revenue Code (LRC) and the forest laws should be suitably modified along with the simplification of felling, transit and trading rules, wherever necessary, to facilitate and motivate individuals and institutions to undertake tree-farming and the growing of tree crops on their own land.

4.10.5 The management plan/working prescriptions should be strictly followed for raising plantations inside the government forest areas.

4.10.6 Irrigated and high input plantation of tree crops should be encouraged for meeting the timber demand of the state. The State Forest Development Corporation, FDC should play a pivotal role in this endeavour.

4.11 Production of Biomass

For the vast majority of the people in the state, the foremost need is for fuelwood, timber, fodder and fibre. The issue of enhanced production and sustainable resource utilization should, therefore, be prioritized in favour of and with due regard to the requirements of the rural people.

4.11.1 The management/working plan prescriptions should guide the production of biomass including timber from the forests of the state. Necessary steps to promote efficient conversion and utilization of timber should be promoted for the maximization of resource use.

4.11.2 Promotion of alternative sources of domestic energy should be taken up on a priority basis to reduce pressures on forests for the supply of fuelwood.

4.12 Forest Based Industries

In consonance with the National Forest Policy, 1988, forest based industries should be encouraged to produce their own raw material through private forestry and to use alternative raw material.

4.12.1 No forest based enterprise, except that at the village or cottage level, should be allowed in future without a proper ecological, cultural and social impact assessment. The fuel, fodder and timber requirements of the local population should not be sacrificed for raw material supplies to such enterprises.

4.12.2 Direct relationship between forest based industry and farmers should be encouraged to meet the raw material requirements of the industry. This industry-farmer collaboration should in no way be allowed to result in diversion of prime agricultural lands and displacement of small and marginal farmers.

4.12.3 The bio-mass resources of the state should not be subsidized to the industry, which should be encouraged, to the extent possible, to use alternative non-forest raw material.

4.12.4 Allotment of land to the industry should be subject to land ceiling and other land laws of the state. Such industry should not in any way be allowed to adversely affect the socio-cultural traditions of the tribals and other communities living in the state.

4.12.5 Appropriate institutional and technological systems should be developed to enable rural artisans to sustain their bio-massbased crafts and enterprise.

4.13 Tribal People and Forests

Having regard to the symbiotic relationship between tribal and forests, a primary task of all agencies responsible for forest management, including the forest department, the Forest Development Corporation and the Minor Forest Produce Federation should be to associate the tribals closely in the protection, regeneration and development of forests as well as to provide gainful employment to people living in and around the forests on following lines.

- Protection, regeneration and non-destructive harvesting of minor forest produce in collaboration with the local people specially tribals, and provision of institutional arrangements for the marketing of such produce;
- Conversion of forest villages into revenue villages;
- Community based schemes for improving the economic status of the tribals;
- Undertaking integrated area development programmes to meet the needs of the tribal economy and to reduce the pressure on the existing forests.

4.14 Forest Extension

Forest conservation programme cannot succeed without the willing support and co-operation of the people. It is essential, to inculcate in the people a direct interest in forests, their development and conservation, and to make them conscious of the value of trees, bio-diversity and nature in general. This can be achieved through the involvement of educational institutions, right from the primary stage.

Farmers and other interested groups should be provided opportunities through different institutions to learn and adopt agri-silvicultural techniques to ensure optimum utilization of their land and water resources. Suitable programmes should be propagated through mass media using audio-visual aids and the extension machinery already existing with the universities and the government departments.

4.14.1 *Promotion of nature-tourism*

Nature tourism or eco-tourism that utilizes the forest scenic spots as well as the opportunities provided by the protected areas for wildlife viewing should be seen as a forest extension activity. This activity should also be promoted as a revenue generating mechanism that can benefit the rural communities through their active involvement in promotion of eco tourism.

4.15 Forestry Education

Forestry should be recognised both as a scientific discipline as well as a profession. Universities and institutions dedicated to the development of forestry education should impart academic education and promote post-graduate research and professional excellence, keeping in view the manpower requirements of the state. Academic and professional qualifications in forestry should be kept in view for the recruitment of the State Forest Service personnel.

4.16 Forestry Research

An increasing recognition of the importance of forests for

environmental stability, as a source of energy, and as a provider of essential requirements and employment to rural poor, calls for scientific forestry research, by adequate strengthening of the research base as well as by setting new priorities for action. The state should promote, aid and co-ordinate research projects, to be undertaken by research organizations like the ICFRE (Indian Council of Forestry Research and Education), and universities, in a transparent and competitive manner. Some broad priority areas of research and development needing special attention in the state are:

- Evolving innovative multi-tier silvicultural systems with integrated ecosystem approach.
- Increasing the productivity of non-wood forest produce per unit of area per unit time by the application of modern scientific, silvicultural and technological methods.
- Revegetation of barren/marginal/waste/mined lands and watershed areas.
- Effective conservation and management of existing natural forest resources.
- Social forestry, farm forestry and agro-forestry.
- Establishment of modern nurseries in all districts by the department as well as by private individuals.
- Emerging areas like the joint forest management and appropriate silvicultural practices; efficient utilization of forest resources; women and tribal empowerment; role of forestry in poverty alleviation; social and livelihood analysis of forest dependent communities; forest policy; cultivation and marketing of medicinal plants; conservation of threatened and endangered species of wild flora and fauna; forest management at landscape level; conservation of bio-cultural values of forests; etc.

4.17 Personnel Management and Capacity Building

Government should aim at enhancing the professional competence status of foresters. It should attract and retain qualified and motivated personnel, and provide conducive working atmosphere, in view of the arduous nature of duties they have to perform, often in remote and inhospitable places.

Capacity building of local people specially the members of VFC (Village Forest Committee), FPC (Forest Protection Committee) and EDC (Eco Development Committee) should form an integral part of the Human Resource Development (HRD) strategy of the state.

4.17.1 The government should ensure sustained availability of trained manpower at every level of the forest department's hierarchy. There should be regular recruitment of forest staff.

4.17.2 Specialized and orientation courses for in-service foresters should be conducted on a regular basis. Management Development programmes incorporating the latest developments in forestry and related disciplines, should be promoted.

4.18 Application of Information Technology into Forestry

4.18.1 *Development of a Comprehensive Forest Database.*
Priority needs to be accorded to developing a comprehensive database regarding the forest resources in the state and to update it on a regular basis. A Forest Information Centre (FIC) should be set up and well provided for in terms of man power as well as computation hardware and software resources.

4.18.2 Use of Geographical Information System (GIS) and Global Positioning System (GPS) in forest management.
Geographical Information System (GIS) and Global Positioning System (GPS) technology has important applications in forestry. A full-fledged GIS centre for rapid adoption of these technologies into the planning, implementation and monitoring of forestry plans and schemes should be established.

4.18.3 *Promotion of Electronic Governance in Forestry*
Electronic governance, which is the application of information technology to the process of governance, has assumed importance in all walks of life. Forest administration should be encouraged to maximize the use of e-technology in all its operations specially that which pertains to public dealing.

4.19 Legal Support and Infrastructure Development

Appropriate and adequate legislative and infrastructure support would be necessary for an effective implementation of this policy.

4.20 Financial Support for Forestry

The objectives of the state policy cannot be achieved without the adequate investment of financial resources on a scale which is appropriate to the need. The state should endeavour to arrange financial resources for the implementation of forestry programmes flowing from this policy document.

4.21 Epilogue

It is envisaged that this policy should usher in forest management programmes in Chhattisgarh, that will respect the thresholds of environmental stability, promote conservation of bio-cultural heritage, and fulfil basic requirements of forest based and forest adjacent tribal and analogous communities.

By Order and in the Name of the
Government of Chhattisgarh,

Ram Prakash,
Special Secretary

APPENDIX III

Approval of Working Plan of Rajnandgaon (Forest Division of Government of India)

भारत सरकार
GOVERMENT OF INDIA
पर्यावरण एवं वन मंत्रालय
MINISTRY OF ENVIRONMENT & FORESTS

No: 12-8 2000 (FOR) 2342

To
The Principal Secretary,
Forest Department,
Government of Chhattisgarh,
D.K. Bhawan,
Raipur-492001

Sub: Approval of Working Plan of Rajnandgaon Forest Division for the period from 1999-2000 to 2008-2009 Written by Shri K.C. Bebarta.

Ref: GOMPs letter No.F-25/9/10-2'2000, dated 26.5.2000.

Sir

With reference to the above mentioned subject I am directed to inform you that after careful examination of the Working Plan of the Rajnandgaon Forest Division, the Central Government hereby conveys its approval to the said working plan in accordance with the powers vested under Forest (Conversion) Act, 1980 subject to the following general and specific conditions:-

A. GENERAL CONDITION:

(i) The orders of Hon'ble Supreme Court in the matter of Godaverman Therumalkpad *Vs* Union of India in W.P. No 202/95 and related Inter Locutory applications shall be strictly adhered to. Any prescription or opinion in variance with the Hon'ble Supreme Court's order is in force or otherwise modified.

(ii) Further, in compliance with Hon'ble Supreme Court's order dated 22.09.2000, the State Government of Chhattisgarh shall ensure that regeneration of forest is commensurate with fellings carried out under this working plan.
No felling shall be carried out without allocating necessary fund for implementation of regeneration operation so as to make regeneration carrying out regeneration operation. No further felling shall be undertaken until the failure shortfall is made up.

(iii) The currency of Working Plan shall be only for a period of 10 i.e. from 1999-2001 to 2008-2009.

(iv) No forest bearing naturally grown trees shall be clear felled for purpose whatsoever.

(v) All kinds of fellings including that of dead, dying and diseased trees and for granting of rights and concessions as well as all illicit fellings should be compiled annually along with the estimation of their stand volumes as per the same volume table used for assessment of growing stock. This report shall be prepared annually Working Circle and Compartment-wise by the territorial DFO and shall be submitted to the C.F. (Working Plan) within two months of the end of control year in a proforma, to be designed by the CF (Working Plan) for this purpose.

Following nature of deviation shall be reported control yearwise to the Regional Office of this Ministry through the Conservator of Forests (Working Plan) every year before the end of September. No further felling of any kind will be undertaken in the next control year unless the following nature of deviations of previous control year have been sanctioned by this office.

(a) Non-implementation of conditions, stipulated in this approval under the heading of "SPECIFIC CONDITIONS".
(b) Non-submission of control forms and deviations statement to the competent authority of the Sate Government and failure to obtain their approval as per the laid down procedures in the State.
(c) Deviation for excess yield/fellings in case the cumulative yield/felling in any Working Circle at the end of any particular year exceeds by more than 10% of the approved yield/fellings upto that year.

B. SPECIFIC CONDITION

1. (i) *Selection-cum-Improvement Working Circle :-*
Felling of trees shall be restricted to 50% of silviculturally available trees.

(ii) *Improvement Working Circle:-*
Under 'D' category felling of silviculturally available trees only shall be carried.

(iii) *RDF Working Circle :-*
Felling of only dead, diseased and dying trees is allowed.

(iv) *Plantation Management Overlapping Working Circle:-*
Thinning of older plantation shall only be carried out after taking written approval of silviculture wing of Forest Department. A detailed mid-term appraisal of success and establishment of plantation shall be carried out in 2004 by the CF Working Plan and the appraisal report shall be submitted to this office. In case of any failure, corrective measures shall have to be prescribed in consultation with the Regional Office of Central Government. It shall be ensured that no area other than artificial plantation or blanks are included in the W.C.

(v) *Rehabilitation of Degraded Forest Working Circle :-*
Commercial exploitation of bamboo shall not be done. National regeneration shall be encouraged alongwith bamboo planting in open patches.

(vi) *Bamboo Overlapping Working Circle :-*
Commercial exploitation shall be strictly done as per the prescription and natural regeneration shall be supplemented with artificial regeneration.

2. Following the direction of Hon'ble Apex Court in their order dated 22.09.2000, a Core Group has been constituted under the Chairmanship of Director General of Forests and Special Secretary for deciding the extent of harvesting that could be permitted under approved Working Plans for ensuring regeneration to be commensurate with fellings. Instruction/ directions of the Central Government in the matter to be issued in future shall be strictly compiled with. Felling to be done by the State Government only after seeking permission from Core Group constituted by MoEF, New Delhi.
3. Felling carried out on forest land after seeking approval of the Central Government under Forest (Conservation) Act, 1980 will not be treated as deviation. However, proposed felling in the affected Working Circle shall be restricted proportionately in the current/following years to compensate this removal.
4. The unauthorized encroachments shall be removed as per the scheme submitted which will be reviewed periodically by the Additional Principal Chief Conservator of Forest (WP) and report/status be submitted to this office. The list of encroachment furnished is not as per the guidelines of Forest (Conservation) Act, 1980, hence not approved.
5. Adequate funds shall be allotted by the State Government for implementation of the prescriptions of Working Plan including boundary demarcation to the Forest Department.
6. The Central Government reserves the right to review, modify, withdraw this approval at any time if any of the conditions of approval are not implemented or relevant modifications in the Working Plan is required so as to keep it in conformity with the order circulars and guidelines issued by the Central Government of the Apex Court under Forest (Conservation) Act, 1980 or any other statute and National Forest Policy.

Yours faithfully

(S. P. Yadav)

Dy. Conservator of Forest (Central)

Glossary

Age Class Distribution: The local occurrence, or proportionate representation, of different Age Class in a forest

Age Class: One of the intervals into which the range of age of trees growing in a forest is divided for classification or use; also the trees falling into such an interval

Agricultural Lands: Fields, pastures, and orchards that are managed to produce food and fibre for people. Agricultural lands provide a source of income and employment, open space for recreational activities, and important habitat for wildlife. Agricultural lands can also be a source of negative environmental impacts (e.g., water pollution from nutrients, eroded soils, and pesticides).

Agro-ecosystem: A dynamic association of crops, pastures, livestock, other flora and fauna, atmosphere, soils, and water. Agro-ecosystems are contained within larger landscapes that include uncultivated land, drainage networks, rural communities, and wildlife.

Alluvium: A general term for clay, silt, sand, gravel, or similar unconsolidated material deposited by a stream or other body of running water.

Annual allowable cut or ACC: The amount of timber that is permitted to be cut annually from a particular area. It is used to ensure a sustainable supply of timber. It averages roughly the amount of new growth each year minus deductions for losses due to fire, insect or diseases.

Anthropogenic: Originating from man, not naturally occurring.

Aquifer: An area of underground soil or rock, both porous and permeable, that allows ground water to easily move. Aquifers are capable of yielding supplies of water and typically consist of unconsolidated deposits or gravel, sand, sandstone, granite, or fractured rock such as limestone. Aquifers can be classified as confined or unconfined.

Atmosphere: The blanket of gases, vapor, and aerosol particles encasing the Earth.

Atmospheric Deposition: The flux (flow) of chemicals and materials from the atmosphere to the Earth's surface. Depending on the chemical or material, "dry" deposition (e.g., by particles) can be less than, equal to, or greater than "wet" deposition (e.g., precipitation).

Available Soil Moisture: The portion of water in a soil that can be readily absorbed by plant roots. It is the amount of water released between *in situ* field capacity and the permanent wilting point.

Biodiversity: Biological diversity, which includes the diversity of species, genes and ecosystem and the evolutionary and functional process which link them.

Biodiversity: The variety and variability among living organisms and the ecosystems in which they occur. Biodiversity includes the number of different items and their relative frequencies; these items are organized at many levels, ranging from complete ecosystems to the biochemical structures that are the molecular basis of heredity. Thus, biodiversity encompasses expressions of the relative abundances of different ecosystems, species, and genes.

Biosphere: The envelope of Earth (including all organic matter, both living and nonliving) and its atmosphere that can support life.

Block: (a) A main territorial division of forest, generally bounded by natural features and (b) bearing a local proper name.

Canopy, Closed: A Canopy in which the individual tree crowns generally touch one another.

Canopy: The cover of branches and foliage formed by the crowns of trees in a wood.

Climate: The average of patterns of a region or locality, based upon long-term statistics, including extremes (such as flood and droughts) and deviations from normals.

Climax : The culmination stage in plant Succession for a given Environment.. Used as an exact synonym of Climax, Climatic by Clements and his school.

Climax Edaphic : A community which differs from the climatic Climax of the area owing to the influence of special soil factors.

Climax, Biotic : A Climax which differs from the climatic Climax of the area owing to the action of Biotic Factor.

Climax, Climatic : A Climax which owes its distinctive characters to Climatic Factors in conjunction with only such biotic influences as plants and animals naturally occurring in the area bring about.

Clump: (*a*) The aggregate of stems issuing from the same root. (b) Rhizome system or Stool with particular reference to bamboo's and the larger grasses.

Community: The assemblage of populations of plants and animals that interact with each other and their environment. The community is shaped by populations and their geographic range, the types of areas they inhabit, species diversity, species interactions, and the flow of energy and nutrients through the community.

Compartment : A territorial unit of a forest permanently defined for purposes of administration, description and record.

Compartment History: A record of all events effecting the forestry of an individual compartment.

Coppice Shoot: A shoot arising from an adventitious bud at the base of a woody plant that has been cut near the ground or burnt back. Sometimes (incorrectly) used to include Root Sucker.

Coppice, Seedling: Coppice Shoot arising from the base of seedling that have been cut or burnt back.

Coppice, To: (*a*) Transitive. (b) To fell trees near ground level with a view to their producing coppice shoots.

Coppice: A crop of Coppice Shoot.

Coupe : A felling area, usually one of an annual series unless otherwise stated. Preferably numbered with Roman numerals, I, II, III, etc.

Criterion: An aspect that is considered important by which sustainable forest management may be assessed. A criterion is accompanied by a set of indicators.

Cropland: An area cultivated to provide various food products.

Customary rights: Rights which result from a long series of habitat or customary actions, constantly repeated, which have, by such repetition and uninterrupted acquiescence, acquired the force of a law within a geographical or sociological unit.

Cut Back, To: To cut flush with the ground. Apart from its application to the cutting of weed and climber in tending, generally, it is also used for the cutting of a stool or any stem in the pole stage and below with the object of obtaining superior coppice growth.

Deciduous: (*a*) Shed naturally : applied to any plant organ or group of organs. (*b*) Also applied to such perennial plants as are normally leafless for some time during the year.

Density, Canopy : The relative completeness of Canopy usually expressed as a decimal coefficient, taking Canopy, Closed as unity. The following classification of canopy density is in vogue:-
Closed when the density is 1.0
Dense when the density is between 0.75 and 1.0.
Thin when the density is between 0.5 and 0.7;and
Open when the density is under 0.5 *Cf.* Density, Crop.

Ecology: The study of the interrelationships between organisms and their natural environment.

Ecosystem: A dynamic and inter-related community of biological organisms and the surrounding environment linked through nutrient cycling and energy flows.

Ecosystem: The most complex level of organization is the ecosystem. An ecosystem includes the plant and animal communities in an area together with the non-living physical environment that supports them. Ecosystems have physically defined boundaries, but they are also dynamic: their boundaries and constituents can change over time. They can import and export materials and energy and thus can interact with and influence other ecosystems. They can also vary widely in size.

Ecotone : The transition zone (or tension belt) between two adjoining community irrespective of their ecological status. Ecotone of an individual organism or taxonomic group as opposed to Synecology which is the ecology of a community.

Enumeration : The counting, singly or together, of individuals of one or more species in a *forest crop* and their classification by species, size, condition, etc. Enumeration may be *complete (total)* or *partial (sample)*; a partial enumeration may by carried out on separate sample plots or in strips or lines.

Erosion: The wearing away of the land surface by running water, wind, ice, or other geological agents, including such processes as gravitational creep. Geological erosion is erosion that occurs naturally over long periods of time. Accelerated erosion is more rapid than normal erosion and results primarily from man's activities. Erosion is further classified by the amount and pattern of soil removal and transport.

Evaporation: The process of returning moisture to the atmosphere. Water on any surface, especially the surfaces of mud-holes, ponds, streams, rivers, lakes, and oceans, is warmed by the sun's heat until it reaches the point at which water turns into the vapour, or gaseous, form. The water vapour then rises into the atmosphere.

Evapotranspiration: The combined loss of water to the atmosphere from land and water surfaces by evaporation and from plants by transpiration.

Evergreen: Never entirely without green foliage, leaves persisting until a new set has appeared.

Extinct: A species of plant or animal that is no longer living.

Felling Cycle: The time which elapses between successive main felling on the same area.

Felling Series. A forest area forming the whole or part of a Working, Circle and delimited so as (1) to distribute Felling and Regeneration to suit local conditions and (2) to maintain or create a normal Age Class Distribution. The yield is calculated separately for each felling series which should have an independent representation of Age Class.

Felling, Final: The removal of the last seed or shelter trees after Regeneration has been effected under a shelter wood system.

Felling, Improvement: The removal or destruction of less valuable trees in a crop in the interest of better growth of the more valuable individual; usually applied to a mixed, Error! Bookmark not defined. It may include thinning of closely stocked groups along with clearing and general assistance to young growth of valuable species. Felling Cycle. The time that elapses between successive Felling Main on the same area.

Forest : (*a*) An area set aside for the production of timber and other forest produce, or maintained under woody vegetation for certain indirect benefits which it provides, e.g., climatic or protective. (*b*) (*Ecological*). A plant Community predominantly of trees and other woody vegetation, usually with a Canopy, Closed. (*c*) (*Legal*). An area of land proclaimed to be a forest under a forest law.

Forest Cover: All lands more than one hactare, with a tree canopy dinsity of more than 10 percent such land not be statutorily notified on farm's area (SFR-2003)

Forest Degradation: A reduction in the biomass, productivity or biodiversity of a forest, and alterations in ecological functions such as habitat, climate regulation and soil and water protection. Sometimes also referred to as a lowering in the quality or integrity of a forest ecosystem. One example is a change in forest composition, which negatively affects the growing stock.

Forest Inventory: The measuring and distributing the quantity and quality of forest crop and many other characteristics of the land area upon which forest crop is growing.

Forest Management Unit (FMU): An area of forest under a single or common system of forest management.

Forest Management: The practical application of the scientific, technical and economic principles of forestry.

Forest Type: A category of Forest defined generally with reference to its geographical location, climatic and edaphic features, composition and condition.

Forest, Coppice: A forest consisting of trees derived mainly form Coppice shoot or Root Sucker.

Forest, High: (*a*) A crop or crops of trees, normally with a closed and high Canopy and usually of seedling origin. (*b*) A forest which originated from seed. (*c*) Sometimes used for a Forest, Closed of tall trees in contrast to Savannah Woodland or Scrub.

Forest, Mixed: A forest composed of tree of two or more species intermingled in the same Canopy; in practice, and by convention, at least 20% of the principal one. The species composing the mixture may be distinguished as *principal, accessory* and *auxiliary*.

Forest, Protected: A legal term for an area subject to limited degree of protection under the provisions of Chapter IV of the Indian Forest Act.

Forest, Reserved: An area so constituted under the Indian Forest Act or other forest law.

Forest: Communities characterized by a predominance of woody vegetation growing more or less closely together, whether 'closed' or 'open' in canopy, and whether fully intact ecologically or modified, fragmented or planted by humankind. Technically forests are defined as ecosystems with a minimum of 10 % tree canopy cover.

Forest: Land with at least 10% of its surface area stocked by trees of any size or formerly having had such trees as cover and not currently built-up or developed for agricultural use. Forests are distinguished by their species of trees and plants, density, and soil type. Forests are also divided into vertical layers: the canopy, the shrub layer, and the forest floor.

Forestry: The theory and practice of all that constitutes the creation, conservation and scientific management of forest and utilization of their resources.

Fragmentation: Refers to a formerly healthy forest that has been divided into many patches, usually due to conversion to agricultural or residential land.

Geographic Information System (GIS): A collection of computer hardware, software, and geographic data designed to capture, store, update, manipulate, analyse, and display geographically referenced data.

Girth Class: One of the intervals into which the range of girth of trees or logs is divided for classification or use; also the trees or logs falling into such an interval.

Girth, Exploitable: The minimum girth at breast height at which trees are considered suitable for conversion.

Grazing: (a) The eating of any kind of standing vegetation by domestic livestock or wild animals. Sometimes limited to the eating of Herbage in contrast with Browsing. (b) Used sometimes as a synonym of Forage.

Gregarious Flowering. The general flowering, within one or a few years, and over considerable area, of all or most of the individuals of certain species that do not flower annually; in some cases followed by the death of the plant e.g. bamboos.

Growing Stock, Normal: The total volume of trees in a fully-stocked forest with a Age Class Distribution, Normal for a given Rotation. Forest, Normal.

Growing Stock: The sum (by number or volume) of the entire tree growing in the Forest or a specified part of it.

Habitat: The locality, site and particular type of local environment occupied by an organism.

Indicator Plant: Any plant which by its presence, increase or decrease, indicates the quality of the site.

Indicator: An indicator is defined as any component of the environment that quantitatively estimates the condition of ecological resources, the magnitude of stress, the exposure of a biological component to stress, or the amount of change in condition.

Indicators: A quantitative, qualitative or descriptive attribute that, when periodically measured or monitored, indicates the direction of change.

Infiltration: The downward entry of water through the soil surface into the soil. Infiltration constitutes the sole source of water to sustain the growth of vegetation and it helps to sustain the ground water supply to wells, springs, and streams. The rate of infiltration is influenced by the physical characteristics of the soil, soil cover (i.e. plants), water content of the soil, soil temperature, and rainfall intensity. The terms infiltration and percolation are often used interchangeably.

Integrated Assessment: An assessment that brings together data about the condition (past, current and future) of ecosystems; the impact of humans and our lifestyles on the ecosystems; and the effects upon humans of the changed environment. Integrated assessments combine the following scientific disciplines: ecology, social sciences, health science, and engineering.

Land Use Plan: A coordinated composite of information, ideas, policies, programes, and activities related to existing and potential uses of land within a given area and frequently the key element in a comprehensive plan for an area under development for public and private land uses, such as residential, commercial, industrial, recreational, and agricultural activities.

Land Use Planning: The process of inventorying and assessing the status, potentials, and limitations of a particular geographic area and its resources, interacting with the populations associated and/ or concerned with the area to determine their needs, wants, and aspirations for the future.

Land Use: The primary or primary and secondary uses of land, such as cropland, woodland, pastureland, etc. The description of a particular land use should convey the dominant character of a geographic area, and thereby establish the types of activities, which are most appropriate and compatible with primary uses.

Land Use: The way land is developed and used in terms of the kinds of anthropogenic activities that occur (e.g., agriculture, residential areas, industrial areas).

Landscape Conditions: The apparent status or characteristics of a landscape unit as measured by one or more landscape indicators.

Landscape Indicator: A characteristic of the environment that is measured to provide evidence of the biological condition of one or more resources at the ecosystem level.

Landscape Management: Involves managing an area or region by integrating a variety of values associated with land that may be ecological, social, cultural or economic.

Landscape Unit: Designed to identify repeating patterns associated with dominant land uses in an area, and defined by the relative proportions of forest, agriculture, and developed (urban) land cover contained in that area.

Landscape: The geographical and ecological integrity and resilience of a particular land areas, not mainly its aesthetic qualities. Landscape is not just a geographical concept, but includes human, cultural, and traditional values that are associated with the land.

Landscape: The spatial pattern or structure of a geographic area (including its biological composition, its physical environment, and its anthropogenic or social patterns) and is designed to identify repeating patterns associated with dominant land uses in an area. The relative proportion of forest, agriculture, and urban land cover contained in the area also defines a landscape as does the interrelationships between them.

Locality, Factors of: The effective biotic climatic and edaphic conditions of a site.

Marking: Selection of trees to be cut or retained in a felling or thinning, the selected trees being indicated by a mark made with a scribe, or with a chisel or Marking Hammer on a Blaze.

Maturity: *(a)* The stage at which a tree is capable of reproducing itself by seed and has attained its full height. *(b)* The stage at which a tree of crop has attained Size, Exploitable or Age, Exploitable.

Model: A representation of reality used to simulate a process, understand a situation, predict an outcome, or analyse a problem. A model is structured as a set of rules and procedures, including spatial modelling tools that relate to locations on the Earth's surface

Non-wood Forest Products: Products for human consumption: food, beverages, medicinal plants, and extracts (e.g. fruits, berries, nuts, honey, game meats, mushrooms, etc.) fodder and forage (grazing, range). Other non-wood products (e.g. cork, resins, tannins, industrial extracts, wool and skins, hunting trophies, Christmas trees, decorative foliage, mosses and ferns, essential and cosmetic oils, etc.) (FAO-FRA–terms and definitions).

Nutrients: Essential inorganic chemicals (e.g., nitrogen and phosphorus) needed by plants for growth. Excessive amounts of nutrients (eutrophication) can lead to degradation of water quality by promoting excessive growth, accumulation, and subsequent decay of plants, especially algae (phytoplankton).

Overwood: The uppermost Storey of a Storeyed High Forest or of any crop in which two or more distinct crown layers occur, either temporarily or permanently; e.g. Seed Bearer over Regeneration, standards over Coppice or a crop of trees over bamboos, underwood.

Percolation: The downward movement of water through soil and rock. The terms infiltration and percolation are often used interchangeably.

Permeability: A measure of how well the spaces in the soil or rock are connected. Used to determine the ease with which rock, soil, or sediment will transmit a fluid.

Plantations: Forest stands established by planting or/and seeding in the process of afforestation or reforestation. They are either:

- Of introduced species (all planted stands) or
- Intensively managed stands or indigenous species, which meet all the following criteria; one or two species at plantation, even age class, regular spacing. (FAO-FRA– terms and definitions).

Plot, Sample: A plot chosen as representative of a larger area. In forestry, sample plots are used mainly in Enumeration surveys and for studies of growth (e.g., preparing Yield Table or determining the effect of thinning treatments on the rate of growth).

Pole: A young tree from the time when the lower branches begin to fall off to the time when the rate of height growth begins to slow down and crown expansion becomes marked.

Policy: A rule or norm usually prescribed by governments to help behaviour or decision making.

Precipitation: Moisture that falls from the atmosphere as rain, snow, sleet, or hail. Precipitation varies in amount, intensity, and form by season and geographic location.

Productivity: The conversion of light energy and carbon dioxide into living organic material.

Quality, Site: A measure of the relative productive capacity of a site for a particular species. The top *height* as it varies with age is generally the basis for classification.

Regeneration Area: The area laid down, normally in a working plan or scheme, to be regenerated in a specified time.

Regeneration Survey: A survey for the assessment of established and unestablished regeneration generally by sample enumeration.

Regeneration, Artificial: The renewal of a forest crop by sowing, planting, or other artificial methods; also the crop so obtained.

Regeneration, gap: A method of artificial regeneration in gaps, occurring or created in a forest, usually for supplementing error! Bookmark not defined.

Regeneration, Natural: The renewal of a forest crop by self-sown seed, or by coppice or root suckers; also the crop so obtained.

Regeneration: Regrowth of the forest following harvesting, either through natural process of seed dispersal and germination of artificially by planting.

Regeneration: The renewal of a forest crop by natural or artificial means; also the new crop so obtained.

River: A stream of water of considerable volume, which travels downhill (from higher altitudes to lower altitudes due to gravity). Rivers carry freshwater to cities and farms, serve as the home to wildlife and fisheries, and provide recreation and natural beauty for people throughout the nation. Rivers are used by humans for irrigation, to transport people and their manufactured products, to produce hydroelectric power, and to provide habitats for animals

Rotation of the Maximum Volume Production: The Rotation that yields the greatest annual quantity of material. It coincides with the age at which the mean Increment, Annual *culminates*.

Rotation: The planned number of years between the Form or Regeneration of a crop and its felling, final. In the case of a Forest Selection, the average age at which a tree is considered mature for felling.

Runoff: The flow of water, usually from precipitation, which is not absorbed into the ground. It flows across the land and eventually runs to stream channels, lakes, oceans, or depressions or low points in the Earth's surface. The characteristics that affect the rate of runoff include rainfall duration and intensity as well as the ground's slope, soil type, and ground cover. Runoff can pick up pollutants from the air and land, carrying them into the streams, lakes, etc.

Sample, Stratified Random : A sample from a stratified population consisting of a random selection of Sampling Unit from each stratum; the sampling fraction need not be the same for all strata.

Silvicultural Operations, Subsidiary: A general term for all operations carried out immediately after the Felling Main; besides Cultural Operations proper, this might include other specific operations.

Silvicultural System: Silvicultural system is a method of silvicultural procedure worked out in accordance with accepted sets of silvicultural principles by which crops constituting forest are tended, harvested and replaced by new crops of distinctive forms.

Silviculture: The art and science of cultivating forest crops.

Spatial Database: A collection of information that contains data on the phenomenon of interest, such as forest condition or stream pollution, and the location of the phenomenon on the Earth's surface.

Spatial Pattern: Generally, the way things are arranged on a map. For example, the pattern of forest patches can be described by their number, size, shape, distance between patches, etc. The spatial pattern exhibited by a map can also be described in terms of its overall texture, complexity, and other indicators.

Spatial Scale: For example, a landscape, forest ecosystem, community, etc.

Species: The species is the fundamental unit of biological organization. Members of a species are genetically similar and normally mate only with other members of the same species, thus preserving that genetic similarity.

Stakeholders: People who are interested in, or affected by, forest management and operations: for example, government agencies, local communities, employees, investors, environmental interest groups, non-governmental organizations (NGO) and general public.

Succession: The gradual replacement of one Community by another in the development of vegetation towards a Climax; succession is "primary" on sites which have previously not borne vegetation and "secondary" after destruction of whole or part of original vegetation.

Sustainable Development: Development that meets the needs of the present without compromising the ability of future generations to meet their own needs. It implies integrating traditional economic concerns with often ignores social and environmental considerations.

Sustainable forest management (SFM): Sustainable forest management is a process of managing forest to achieve one or more clearly specified objectives of management with regard to the production of a continuous flows of desired forest products and services without undue reduction of its inherent values and future productivity and without undue undesirable effects on the physical and social environment. A process, rather than a fixed state. As such, universal prescriptions are over simplistic. Elements may include participatory and equitable approaches to decision-making geared towards maintaining ecosystem and landscape functions while also meeting broad economic, social and cultural needs. The management context is all-important.

Sustained Yield: Production of forest products on a perpetual basis, ensuring that the rate of removal of forest products does not exceed the rate of replacement over the long term.

System, Improvement Felling: A method of treatment involving essential the removal of inferior growing stock and tending the better elements of the crop, usually in an irregular forest. It is not a silvicultural system, as it does not aim at regeneration.

Taxa: A taxonomic unit. A taxonomic group of any rank, including all the subordinate groups. Any group of organisms, populations, or taxa considered to be sufficiently distinct from other such groups to be treated as a separate unit.

Tending : Generally an operation carried out for the benefit of a forest crop, at any stage of its life; essentially covers operations on the crop itself and on competing vegetation, e.g., Weeding, Cleaning, Thinning, and even improvement Felling; also Pruning, to climber cutting and girdle, of unwanted growth, but not felling regeneration, nor ground operations like soil working, drainage, irrigation and burning controlled.

Thinning: A felling made in an immature stand for the purpose of improving the growth and form of the trees that remain, without permanently breaking the canopy.

Topographic: The physical features of a surface area including relative elevations and the position of natural and man-made (anthropogenic) features.

Watershed: The entire area of land whose runoff of water, sediments, and dissolved materials (e.g., nutrients, contaminants) drain into a river, lake, estuary, or ocean.

Working Plan: A written scheme of management aiming at continuity of policy and action and controlling the treatment of a forest. The instrument of forest management.

Working Scheme: A term used in some countries to denote a plan of management not so detailed or so nicely calculated as a Working Plan.

Working, Circle: A forest area (forming the whole or part of a working plan area) organized with a particular object, and under one silviculture system and one set of working plan prescriptions. In certain circumstances working circles may overlap.

Yield : The volume or number of stems that can be removed annually or periodically, or the area over which felling may pass annually or periodically, consistent with the attainment of the objects of management.

Yield Regulation: A technique for calculating and controlling the harvest level to ensure that sustained yield is implemented.

Yield, Sustained: (a) The material that a forest can yield annually (or periodically) in perpetuity. *(b)* As applied to a policy, method or plan of management (sustained yield management), it implies continuous production with the aim of achieving, at the earliest practical time at the highest practical level an approximate balance between net growth and harvest, either by annual or somewhat longer period.

Bibliography

ACC Task Force on rural development (1978). Report on the Third meeting of the working group on programme harmonization, Rome.

Ahmed, E., Dreze, J. Hills and Sen, A. (Eds) (1991): *Social security in developing countries*, Clrendon Press, Oxford.

Amstein, S. (1969). 'A Ladder of Citizen's participation in USA', *Journal of American Institute of Planners:* July, pp. 216-24.

Anon : India 1995 A Reference Annual ; Ministry of Information and Broadcasting: Government of India.

Anon: District Statistical Book , District Rajnandgaon, Government of Madhya Pradesh, Bhopal.

Anon, (1953). Glossary of Technical Terms. British Commonwealth Forest Terminology (BCCFT) (modif) ; Empire Forestry Research Association, England.

Anon, (1983). Guidelines for Preparation of Working Plan and Felling in Forests. Ministry of Environment and Forests. Govt. of India.

Anon, (1988). National Forest Policy : Ministry of Forests and Environment, Govt. of India.

Anon, (1983). Abridged Glossary of Technical Terms. Forest Research Institute, Dehradun, India.

Anon, (1997). Rajnandgaon 1997 : District Statistical Book ; Office of District Magistrate, Rajnandgaon. Govt. of Madhya Pradesh.

Arnold, J.E.M.(1997) *Framing the issues, Farms, trees and farmers, response to agricultural intensification*. pp. 10. Earthscan Publication Ltd. London.

Arnstein, S. (1969). 'A ladder of Citizen's participation in USA', *Journal of American Institute of Planners* : July, pp. 216-24.

Bazzaz, F.A. (1983). Characteristics of population in relation to disturbance in natural and more modified ecosystems. pp. 259-75 in H.A. Mooney and M. Godron (eds). *Disturbance and Ecosystems: Components of response: Ecological Studies*, pp. 44. Springer-Verlag, Berlin.

Bebarta, Kailash Ch., (1999). *Teak: Ecology, Silviculture, Management and Profitability*, International Book Publishers and Distributors, Dehradun.

Bebarta, Kailash Ch., (1999). *Working Plan for Rajnandgaon Forest Division*. M.P. Forest Department. Bhopal.

Bebarta, Kailash Ch., (2004). *Forest Resources and Sustainable Development,* Concept Publishing Company, New Delhi=110059.

Bruth-Davy, J. (1938). The classification of tropical woody vegetation types. Imperial Forestry Institute. paper no. 13.

Chambers, R.(1983). *Rural Development, Putting the last first.* Longman. London.

Champion, G . Harry and Seth, S.K. (1964). *General Silviculture for India.* Forest Research Institute. Dehradun.

Champion, G. Harry and Seth, S.K. (1968). *General Silviculture for India.* Forest Research Institute. Dehradun.

Clements, F.E. (1916). *Plant Succession.* Washington, D.C: Carnegie Institution of Washington.

Cohen, John and Norman Uphoff (1977). *Rural Development Participation: Concepts and Measures for Project Design, Implementation, and Evaluation.* Ithaka, Cornell University.

Davis, K.P. (1966). *Forest Management: Regulation and Valuation.* 2nd ed. McGraw-Hill, New York. 519 pp.

D'Arcy. W.E. (1910). *Forest Working Plan,* Govt. Press.

DBB (1994). *Evaluating Sustainable Forest Management: Report of the Working group of experts on Sustainable Forest Management,* Den Haag, The Netharlands.

Economic Commission of Latin America (1973). Popular participation in Development's *Community Development Journal,* Vol. 8, No. 2: pp. 77-93 Oxford.

FAO (1991). Sustainable Management of Tropical Forests. Anex F. In rep. Committee on Forest Development in the Tropics. Tenth Sessions, 10-13. December. 1991, Rome.

FAO (1993), *The challenge of sustainable forest management: What future for the Worlds' forest*? Rome.

FAO (1995), Forest Resources Assessment 1990: Global Systems. FAO Forestry Paper. Food and Agricultural Regeneration of the United Nations, Rome.

FAO (1982). Participation in Forestry, Paper presented by Siteke Kupersus, Department of Forest Management, Agriculture University, Wageninen.

FAO (1997), State of Worlds Forests 1997, Rome, FAO.

Gale, R.P. and Cordray, S.M. (1991). What should forests sustain? Eight answers. *Journal of Forestry* 89(5), 31-36.

Getbig, I. and Sonke, Schmidt (1992): 'Social Security and the Poor: An introduction', Rethinking Social Security, Asian and Pacific Development Centre.

Griffin, Ricky. (1990). *Management.* 3rd Edition. Houghton Mifflin Company, Boston.

Harrison, S. and Fahrig, L. (1995). Landscape pattern and population conservation. pp. 293-308 in L. Hansson, L. Fahrig and G. Marriam (eds). *Mosaic Landscapes and Ecological Process.* Champman & Hall, London.

Hart, R. (1992). *Children's Participation from Tokenism to Citizenship.* UNICEF.

Heuveldop, J. (1994). *Assessment of sustainable tropical forest management.* Mitteilungen der Bundesforschungsanstalt fur Forst- und Holzwirtschaft, Hamburg, no.178.

Hunter, M. L., Jr (1990). *Wildlife, Forests and Forestry.* Prentice Hall, Englewood Cliffs, New Jersey.

International Labour Organisation (1998): 'Introduction to Social Security', p. 3.

International Tropical Timber Organization (ITTO) (1998). *Criteria and Indicators for Sustainable Forest Management of Natural Tropical Forests.* Yokahama, Japan.

ITTO (1992), *Criteria for the measurement of sustainable tropical forest management.* Policy Development series no.3, ITTO, Yokahama, Japan.

Lankly, J.P. (1982). Tropical Forest resources, FAO Forestry paper 30. Rome.

Leuschner, W.A. (1984). *Introduction to forest resource management.* Wiley, New York. 298pp.

Nair, P.K.R (1984) *Soil Productivity aspects of agro-forestry.* International Council for Research in Agroforestry. Nairobi.

Panwar, H.S. (1992), Eco-development: An Integrated Approach to sustainable development for people and Protected Areas in India. Paper presented in the IV World Congress on National Parks and Protected Areas. Caracus. Venezuela.

Paul R. Ehrlich, Anne H. Ehrlich and John P. Hodren (1977). *Eco-science: Population, Resources Environment* (W.H. Freeman, Sanfransisco, 197 to 149) World Resource Institute and

International Institute for Environment and Development, World Resources 1986 (Basic Books, New York, 1986) p. 5.

Paul, S. (1987). Community participation in Development Projects. Discussion papers No 6 . World Bank, Washington DC.

Picket, S.T.A. and White P.S.(1985). Patch Dynamics : a synthesis, In *The Ecology of Natural Disturbance and Patch Dynamics*, (ed.) S.T.A. Picket and P.S. White, pp. 371-384. Orlando, FL : Academic Press.

Picket, S.T.A. and Ostfield, R.S. (1995). The shifting paradigm in ecology. In *A New Century for Natural Resource Management*, (ed.) R.L. Knight and S.F. Bates, pp. 261-279. Washington , DC: Island Press.

Poore, D. (1993). Criteria for sustainable development for forests. *CSCE Technical Bulletin* no. 1, 71-84.

Pretty, J.N. (1994). 'Alternative Systems of Inquiry for Sustainable Agriculture', *IDS Bulletin* 25: vol. 2 pp. 37-48. Brighton, IDS, University of Sussex.

Rapport, D.J. (1989). What constitutes ecosystem health ? *Perspsecitves in Biology and Medicine* 33: 121-32.

Ruthernberg, H. (1980). *Farming systems in the tropics*. Clarendon Press, Oxford.

Sawarkar, V.B. (1995). *A Manual for planning wildlife planning in the protected areas and managed forests*. Wildlife Institute of India, Dehra Dun.

Sdjo, R.A. (1996). Towards and operational approach to public foret management. *Journal of Forestry* 94/8:24-7.

Spurr, S.H. (1979). Silviculture. *Scientific American* 240: 76-82, 87-91.

Spurr, S.H. and Vaux, H.J. (1976). Timber, Biological and Economic Potential Science, 191: 752-6.

Subramanya, R. K. A. (1994). *Social Security in Developing Countries*, Har-Anand Publishers, New Delhi.

Turner, M.G., Gardner, R.H. and O'Neill, R.V. (1995). Ecological Dynamics at broad scales. *Bioscience: Supplement* S-29 to S-35.

United Nations (1993), Report of the United Nations Conference of Environment and Development, Rio de Janiero, 3-14 June 1992, Vol.1, New York: United Nations.

UNEP (1991), Register Of International Treaties And Other Arrangement In The Field Of Environment. Nairobi: United Nations Environment Programme.UNEP/GC16/Inf.4.

Upton, C. and Bass, S. (1995), *The Forest Certification Handbook.* Earth Scan Publication, London.

WCED (World Commission on Environment and Development - The Brundtland Commission) (1987). *Our Common Future. Report of the WCED*, Oxford. Oxford University Press.

Woodwell, G.M. (1970), Effect of Pollution on the structure and Physiology of ecosystems science, 168: 429-433.

World Bank (1978). Forestry Sector Paper: World Bank. Washington. D.C.

World Bank (2001). *World Development Report 2000/2001 : Attacking Poverty.* Oxford, UK, Oxford University Press.

WRI (2000) World Resources 2000 – 2001 : People and ecosystems : The fraying web of life. Washington, DC.

Wunder, S, (2001) Forestry Elevation and Tropical Forests - What Scope of Synergies. *World Development*, 29 : 1817-1833.

Index